◀ MOTORIZED OBSESSIONS ▶

Life, Liberty,
and the Small-Bore Engine

PAUL R. JOSEPHSON

The Johns Hopkins University Press
Baltimore

Printed in the United States of America on acid-free paper
2 4 6 8 9 7 5 3 1

The Johns Hopkins University Press
2715 North Charles Street
Baltimore, Maryland 21218-4363
www.press.jhu.edu

Library of Congress Cataloging-in-Publication Data
Josephson, Paul R.
Motorized obsessions: life, liberty, and the small-bore engine /
Paul R. Josephson.
p. cm.
Includes bibliographical references and index.
ISBN-13: 978-0-8018-8641-6 (hardcover : alk. paper)
ISBN-10: 0-8018-8641-4 (hardcover : alk. paper)
1. Small gasoline engines. 2. Internal combustion engines—History.
I. Title.
TJ790.J67 2007
621.43—dc22 2006035459

A catalog record for this book is available from the British Library.

CONTENTS

PREFACE

I began this book out of curiosity about the place of internal combustion engines in American society. I had in mind not the large engines in our automobiles but the small-bore engines that power gardening equipment and recreational machines.

Like most Americans, I bought a lawn mower soon after buying a house, and I took pride in creating a green carpet of lawn. Eventually I also bought a line trimmer to create perfect edges and to save time weeding. If you live in a neighborhood with lawns of any size, small or large, urban, suburban, or rural, you too are bound to see neighbors using gardening machines of various sorts to assist them in keeping their lawns beautiful. Some of them also buy ride-on mowers with snowplow attachments to remove ice and snow in the winter. Others purchase cultivators. And so on.

I began to wonder how long these gardening and snow removal machines had been around, how they had evolved, how much Americans spend on them, and when local, state, and national governments began regulating them for safety, emissions, and noise.

Moving to a small town in central Maine some years ago, I also noted small-bore engines increasingly used to power recreational vehicles: snowmobiles, jet skis (also known as personal watercraft, or PWCs), and those vehicles variously referred to as all-terrain, off-road, and off-highway vehicles (ATVs, ORVs, and OHVs; in what follows I generally call them ATVs). I considered their ubiquity in the forests, along the Atlantic coast, and on the myriad lakes created during the last ice age here in Maine. As a cross-country skier and a long-distance runner who has shared trails with snowmobiles and ATVs, I pondered how the machine recreationists' clubs organized their trail systems. I was amazed at the numbers of machine recreationists, their seasonal ubiquity, the joy they derive from their hobby, and above all the energy they devote to organizing local clubs. My curiosity led me to this history of recreational machines and their central place in work and leisure in postwar America.

ATVs, snowmobiles, and jet skis are beautiful machines. They come in a variety of jet-age colors. Their plastic cowlings indicate both aerodynamic efficiency and high craftsmanship. Over the years their manufacturers have perfected their operation through improved suspension systems, quieter and more efficient engines, and better transmissions. These high-powered, highly maneuverable machines can, and do, go everywhere. They represent both world-class high technology and, at their most basic level, the simple grace of rudimentary machines: an engine for propulsion, a steering mechanism, suspension, and seats (although contemporary machines come with a variety of attachments and options like automobiles and such comforts as heated handgrips for snowmobiles). These machines serve not only recreational ends but also utilitarian ones. Weekend gardeners attach winches, cultivators, and other devices to help them shift detritus into different piles and configurations. Utility companies employ them. Many hunters and fishers swear by them. When needed in emergencies, recreational machines serve law enforcement officers, game wardens, medical technicians, or private citizens needing to move sandbags to protect their homes from floodwaters. The ATV has gone to Iraq for use by American soldiers.

Over the past three years I have had the opportunity to discuss the history of recreational machines with recreationists from all walks of life. Members of snowmobile, ATV, and jet ski clubs have talked with me and given me the opportunity to comprehend the joy of playing at high speed on these beautiful, fast, sleek machines. I now understand better their interests, motivations, and concerns, and why they react as they do to efforts to regulate their machines in terms of safety, emissions, noise, or access. Club members are almost without exception family oriented and interested in safe and responsible operation. A number of my acquaintances have machines and love them for play as well as for utility. I have had the honor to be the guest of the Polaris factory in Roseau, Minnesota, and the Bombardier factory in Valcourt, Quebec. From workers on the assembly lines to the managers and owners, all take pride in the safe, exciting machines they build. I have talked with retailers, with environmentalists, and with state officials responsible for licensing and fees and for working with clubs to establish trails. I have spoken with people involved either directly or indirectly in recreational machining, in the medical profession, in consumer product safety organizations, and in state and federal agencies.

Recreational machining has become a social and cultural institution that stretches across North America, from coast to coast and from south to north. While some see recreational machining as a "redneck" sport, it is anything but. It includes Canadians and Americans of all classes and income levels, although not so many people of color. Every weekend North Americans hitch trailers to their cars and SUVs, load them up with jet skis, ATVs, and snowmobiles, fill them with the proper equipment of helmets, personal flotation devices, insignia shirts, pants, gloves, and boots, and spend hours and hours driving to parks, forests, fields, grasslands, lakes, rivers, and coasts in search of recreational nirvana.

To write this story, I have consulted the growing documentary record of recreational machine history. I have read brochures, newspapers, and other documents from the manufacturers; such government documents as Environmental Protection Agency reports, regulatory decisions, executive orders, product recalls, and correspondence between and among officials and other concerned citizens; such medical journals as the *Journal of Trauma* and the *New England Journal of Medicine;* and scientific journals in which biologists, geologists, and others attempt to evaluate the impact of recreational machines on various ecosystems. After all, Americans and Canadians now own millions of machines and use them wherever they are permitted to use them in increasing numbers. What is their environmental impact?

My research revealed a surprising number of environmental and public health issues concerning ATVs, snowmobiles, and jet skis that have not yet been treated systematically in one place. My simple curiosity about recreational machines gave way to concern about their environmental and public health effects, and also about our nation's failure to reach reasonable standards for regulating these machines in terms of pollution, safety, and access policies. President Richard Nixon in 1972 issued an executive order stipulating study of recreational machines and policies concerning their use on federal land units, but federal officials have yet to achieve that goal over thirty years later. In fact, many managers in the National Forest Service, National Park Service, Fish and Wildlife Service, and Bureau of Land Management remain ignorant of the impact of recreational machines and have done little of what the law requires of them to limit that impact. They have not made any systematic study of the widely available data relating to that impact, nor have they responded promptly to the problems these data indicate. It has fallen to state governments—belatedly and with little funding—to work out with machine recreationists, their clubs, and manufacturers how and to what extent to regulate machine use; how to offer (and whether to require) training programs; whether (and how) to implement licensing requirements and fees; whether to require helmet use; and how to establish and maintain the areas (usually trails) where recreationists may tread.

For their part, manufacturers have apparently tried to limit, postpone, or avoid strict, formal regulation by working with government bodies toward voluntary standards. For example, in 1996 the federal government established standards to improve small-engine efficiency. The Environmental Protection Agency established those standards because the small-bore engines used in gardening and recreational machines had become the nation's major single source of air pollution. Yet a voluntary, cooperative effort to manufacture engines that met reasonable and achievable standards suddenly seemed to give way to an all-out battle to weaken and postpone them, a battle apparently conducted with the acquiescence of the administration of George W. Bush. Many manufacturers maintained that they could not meet the new standards without great cost and loss of business, and that new safety standards such as seatbelts or roll bars would raise the cost to consumers, "dumb down" the machines, and ruin the riding experience. These assertions certainly merit consideration. However, EPA studies indicate that the efficiencies to be achieved from cleaner engines would enable consumers to recoup any higher costs of the new machines in lower gasoline, oil, and maintenance costs over the lifetime of the vehicle. Further, manufacturers clearly have the technological acumen to meet the standards, and many quickly—and with great fanfare—did. And, finally, if the standards were to be met by the second decade of the twenty-first century, it would still be another fifteen years before older, more heavily polluting vehicles had been retired. Further delay is therefore both unnecessary and quite dangerous for public health and the environment.

The millions of ATVs, snowmobiles, and jet skis that appear every weekend in all ecosystems—in deserts and other arid climes, in forests and grasslands, in lakes, rivers, and streams, in estuaries and riparian sites, and in the ocean—have another serious impact that must be addressed from both historical and policy perspectives. I offer in this book a historical perspective in the hopes of contributing to a discussion of the difficult political decisions that must be made if traditional, nonmotorized recreationists and motorized recreationists are to share parks, forests, and fields, and if fragile ecosystems already under duress are to be preserved. Of course, most of us recognize that no "pristine" landscape exists. Humans have been everywhere and have left an impact, small or large. Many people, both machine recreationists and not, simply do not recognize that any use of an ecosystem, even such nonmotorized activities as hiking or kayaking, will leave some impact. The biological literature shows that machines have an immediate, extensive, and almost irreversible impact. The damage affects both flora and fauna and results in a loss of biodiversity. Many people believe that deserts are devoid of flora and fauna. But deserts too are vital ecosystems that must be treated with greater care than they have been, and desertification has accelerated for a variety of reasons, one of which is excessive ATV use.

Since the late nineteenth century, in an effort to preserve some semblance of unspoiled nature, Americans have established national parks, monuments, and wilderness areas. Unfortunately, recreational machines have encroached upon and frequently damaged ecosystems in these areas. There are now thousands of miles where machines may legally tread, but for many operators these miles are insufficient, and in a number of parks machine recreationists have arbitrarily added new, illegal trails, contributing to the fracturing of ecosystems. At other parks, notably at Yellowstone National Park, local businesspeople in towns near the entrances have pushed to maintain or increase access for such recreational machine users as snowmobilers because of the importance of the sport to their livelihoods. Given the finite amount of land that might be designated wilderness, the millions of acres already accessible to machine recreationists, and the environmental damage that machines inflict in a single pass, it seems an inescapable conclusion that more lands must be designated off-use to machines. What damage would it do to close such places as Yellowstone and other parks to recreational machines, especially given the availability of other places to use recreational machines and the needs of recreationists who do not use machines?

Snowmobile, ATV, and jet ski clubs have been quite successful in encouraging responsible use to protect the environment. Machine recreation also has a significant economic impact: it produces employment, particularly in the tourist sector. But my research also shows that, given the nature of recreational machines, use will always lead to environmental degradation, and too many operators fail to consider the cost of this degradation. Studies that consider economic impact never calculate the economic costs of loss of wilderness; they focus on job creation. What of the law enforcement costs, the medical costs, the costs of loss of life or livelihood, the costs of degradation of property? Since the 1970s both short-term and long-term studies have documented the growing impact of degradation.

Clubs and manufacturers have attempted to limit degradation by encouraging recreationists to "Tread Lightly!" The Tread Lightly campaign, dating to the late 1980s, grew out of a government-industry effort according to which manufacturers' advertisements would depict examples of "responsible use," uses having very light impact on nature: no spinning tires, no uproarious treks through the mud, no churning water. But the Tread Lightly campaign has been co-opted by industry. Advertisements for recreational machines rarely show responsible use. They show big machines moving at high speed through dirt, mud, and water. That indeed may be what these machines are for, but it is misleading to claim that "churning" is compatible with Treading Lightly. It is impossible to Tread Lightly anywhere on a 600-pound high-speed machine with huge knobby wheels. To make matters worse, as acknowledged by club members, industry spokesmen, and government officials, irresponsible users share the same ecosystems. And if responsible users are challenged to Tread Lightly, imagine what damage even a few irresponsible machine recreationists can wreak. Snowmobile clubs have been more successful than ATV clubs in encouraging their members to avoid trespassing on private property, to clean up after themselves, and to maintain safe trails. Now, however, millions and millions of mass-produced machines fill the nation from shore to shore, and their numbers simply make it impossible to Tread Lightly.

Finally, we must be cognizant of the public health ramifications of machine recreation. Who doesn't enjoy the thrill of riding on a brand-new ATV, jet ski, or snowmobile? And what of their important uses on the farm, in maintaining power lines, logging, and so on? Yet the medical literature, reports of the Consumer Product Safety Commission and Centers for Disease Control, and reports of the National Transportation Safety Board and U.S. Coast Guard reveal a growing crisis. We may lament government interference in our private choices, for example listening to our phone conversations, reading our bank records, or requiring us to wear seatbelts or helmets. However, a serious discussion must occur regarding the historical failure to treat recreational machine injury and mortality rates as a sign of public health crisis. Literally hundreds of people die each year from using these machines, and tens of thousands are injured severely enough to be hospitalized. The injuries include internal bleeding, broken bones, and lacerations, in many cases resulting in paraplegia or quadriplegia. The injuries result in days and weeks of lost productivity in the *billions* of dollars, medical costs annually in the *billions* of dollars. Worst of all, children suffer disproportionately. Parents must be aware of the dangers of machine recreation. As for adults (and as for seatbelt use in automobiles and helmet use with motorcycles), should operators of recreational machines be required to use helmets and other safety equipment? Should this be left to the states to decide? Should manufacturers build roll bars, seek improvements in stability by lowering center of gravity and developing industry-wide safety standards? Should seatbelts be required on machines that go over 50, 60, 70 miles per hour?

I urge readers to consider the evidence of environmental and public health costs along with the evidence that machine recreation is clearly fun, exciting, and an important contribution to local economies. There is no denying the thrill of taking an off-road vehicle over sand dunes, of speeding across frozen lakes on a snowmobile, of ATVing through streams and woods, or of jumping waves on a jet ski. I therefore think it especially important that machine recreationists read this book in pursuit of voluntary and cooperative efforts to achieve greater public safety in the use of their machines. Otherwise the future may bring severe restrictions on the operation of ATVs, snowmobiles, and personal watercraft.

I would like to thank Bob Brugger for his advice and suggestions; Rudi Volti and Bob Post, who read parts of the book and offered important criticism; the anonymous reviewer who suggested vital changes after a close, collegial, and careful reading; Steve Saunders, Jon Chapin, Erik Seastead, Dawn Ego, and Rosalea Kimball, who heard about this book on long training runs over the past three years; the kind people of Roseau, Minnesota, and Valcourt, Quebec; Mitchell Johnson at Polaris; Josee Petit at BRP; Paul Jacques, Maine's deputy commissioner of parks and recreation; Mary Yates for copy editing; Carrie Ngo, Miriam Trotschka, Nina Koroleva, Kaitlin McKafferty, and Courtney Kubilis, who helped with research; Alice Burden and John Michel, who commented on some of my ideas; students in my "Luddites Ranting" history-of-technology course; Jess Laniewski, who prepared the index; Lenny Reich, who has written about the snowmobile; and my colleagues in the history department at Colby College, who have made it such a delightful place to work. I know that all of these people intend to ride recreational vehicles safely and responsibly.

MOTORIZED OBSESSIONS

‹ 1 ›

FORDISM AND OUTDOOR RECREATION

THE PINE FORESTS of northern Michigan, the grassy plains of Minnesota, the tidal basins and estuaries of the Florida peninsula, the lakes of New England, the sand dunes of southern California, the arid Redrock Wilderness of Utah, the forests, parks, and wilderness areas under federal, state, municipal, and private ownership—where have snowmobiles, all-terrain vehicles (ATVs), and personal watercraft not been?

I have been unable to calculate exactly how many internal combustion engines Americans currently own and operate. There may be 500 million of them, moving 225 million automobiles and light trucks; pushing golf carts, mopeds, and motorcycles; powering 200 million lawn mowers, snowblowers, chain saws, tractors, trimmers, shredders, grinders, blowers, tillers, aerators, de-thatchers, spikers, pluggers, power brooms, sod-cutters, spreaders, seeders, and other equipment; and enabling operators of some 15 million personal watercraft, snowmobiles, and off-road vehicles to move quickly far from the beaten path. This book is a history of recreational vehicles powered by small-bore internal combustion engines, their manufacturers, owners, and clubs, and their social and environmental impact.

Recreational vehicles have become a fixture of the American lifestyle. Early models appeared at the beginning of the twentieth century along with early models of automobiles. The high-powered, high-speed modern versions came out of tinkerers' garages and fabrication plants after World War II. In the 1960s more and more Americans directed their disposable incomes, and their automobiles, to vacation lands to see the great beauty of the nation, to visit national parks, and ultimately to go off road in pursuit of pristine nature. They used their ubiquitous automobiles and pickup trucks to tow or ferry their ubiquitous snowmobiles, off-road vehicles (ORVs), and jet skis to vacation lands, neighborhood parks, or club-organized and state-funded trails. In the 1970s and 1980s production of snowmobiles, ATVs, such other ORVs as dune buggies and dirt bikes, and personal watercraft (PWCs, also known as jet skis, the name of the Yamaha model), expanded rapidly. On his way out the door of the White House (but not to ride an ATV), President Richard Nixon issued an executive order requiring that managers of federal land "units" (large tracts of forest, park, wilderness, and so on) evaluate whether recreational machines placed undue stress on America's great natural resources. President Jimmy Carter followed up with another executive order along these lines. We still await that evaluation, while recreational machines have grown in number and reach.

Recreational machines occupy a central place in American social, political, and economic life. They represent Fordism in recreation: the mass production of machines to enable more and more people to recreate in forests, fields, plains, on lakes, rivers, and coastal waters, in mud and in arid climes, at any time and any season. Recreational machines were a logical outgrowth of excess engine production capacity in postwar factories, the opening of new markets, cheap oil and gas, and the desire of such inventors and tinkerers as David Johnson and Ed Hinteen of Minnesota and J. Armand Bombardier of Quebec Province to create utility and sport machines that could take people off the beaten path. They saw meaning and purpose in their inventions: the snowmobile ended winter isolation, it facilitated emergency response in the dead of winter, and it enabled winter travelers to see new vistas.

Hinteen, Johnson, Bombardier, and several others soon established small factories to meet growing demand for their vehicles. Johnson's efforts grew into the Polaris Factory in Roseau, Minnesota. Bombardier's efforts became the massive corporation of the same name in Valcourt, Quebec. The communities that surround the Bombardier and Polaris facilities remain, in many ways, small towns. Valcourt has twenty-four hundred residents, Roseau nearly twenty-eight hundred. Of course, the factories are the major employers in the towns. A sense of pride in the factory and in the objects of labor pervades the community. Workers feel at ease making suggestions for design improvements and improvements in production processes. Unions have been unable to establish themselves, for the workers see no great need for them. After all, the owners never shut down to move to a big city, to the outskirts of Minneapolis or Montreal. Modern, safe, well-lit and well-ventilated, spacious, and efficient assembly lines see high-quality recreational vehicles produced day in, day out by conscientious, proud workers.

Even if the psychology of production remains small town in many ways, the experience of recreation does not. The mass production of vehicles has turned recreation into something large scale and industrial, extending from factory to backyards and garages, from trailers and hitches to parks and trails, from distributors and gas stations to clubs and lobbying organizations for manufacturers, environmentalists, and others. Recreational vehicles reflect a series of paradoxes. They were built to meet utilitarian needs but now serve recreation. They came out of small garages, each vehicle slightly different, but now are mass-produced in modern facilities. They were intended to help people enjoy nature, but because of their speed and power they have overwhelmed nature. The numbers of recreational machines grew from hundreds to millions over twenty years. Attention to the air, water, and noise pollution that often accompanied the relatively inefficient but powerful small-bore engines lagged behind growth in numbers of vehicles. The stewards of our natural resources were slow to recognize the damage the machine operators often cause to ecosystems and wildlife. Regulators and managers assumed that recreational machines were no different from automobiles; after all, they are mass-produced on assembly lines using many of the same processes. Over time, they realized that the environmental and public health costs of recreational machine use required some action. At the same time, within snowmobile, ATV, and PWC communities, operators have banded together in clubs to promote responsible operation. A question remains what responsible operation of 700-pound machines that can reach 90 mph means.

Compared with other motorized devices—and certainly with other forms of recreation—recreational machines often are faster, more versatile, and more powerful, and this makes them more risky to operate. Because of these risks, physicians, state and federal officials, members of nongovernmental organizations (NGOs) and agencies concerned with these issues, and club members have sought safety improvements in design and operation of the machines, especially since children have been disproportionately the victims of accidents. The manufacturers maintain that the best solution to concerns about safety is for operators to follow operating instructions in manuals, warning labels, and training programs. While public health officials indicate growing concern over the human toll of machine recreation, however, virtually all owners swear by them for their thrilling rides and for the ability to visit what they take to be pristine nature. Americans have come to believe that recreation must consist of one part high-powered $10,000 machine, one part $1,000 in accessories, and one part $2,000 in trailer equipment. They worry about overregulation of their expensive machines. The embrace of recreational machines suggests a manifestation of the fearlessness of Americans in the face of any frontier, their love of nature, their enjoyment of engines and speed, and of their innate mistrust of government interference in private activities. So the recreationist goes—millions of them go—to the top of hills, through forests, through cascading streams, and across waves with 100-, 200-, 250-hp machines. Are there limits, or ought there to be limits, on machine use?

A recent *ATV Action* magazine indicates how embedded recreational machines have become in the American lifestyle. Its editors published a review of the 2006 Dinli "Cobia 50" children's youth quad (four-wheeled ATV). The accompanying photos show a tiny boy in full battle regalia (gloves, reinforced boots, helmet, goggles, and so on) jumping, riding, and standing proudly near his new machine. The Cobia 50 is intended for six- to eight-year-old riders: voluntary industry standards permit sales of ATVs with engines under 50 cc displacement for use by children younger than eight, and under 90 cc for children younger than sixteen. The Cobia 50, at 49 cc, is no easy ride. The editors note that the machine lacks "meaningful suspension" and "does without engine skids for truly off-road excursions." The six-year-old test driver handled the machine with confidence. But he'll need skill on top of confidence, because "there's a steel tube that runs between the wheels that pivots in the center. It will react slightly to a bump to one wheel, but if the bump hits both wheels simultaneously, there's no suspension action at all." Also, the rear disc brake is mechanical, not hydraulic, meaning that it is relatively weak and in constant need of adjustment. So don't expect the machine, or the young rider, to do well if the terrain is more challenging than a dirt road or a parking lot.[1] A little boy certainly cannot understand his own mortality or the limits to his ability to handle a machine capable of reaching 45 mph. He may not even be able to read the operator's manual or the labels affixed to the ATV warning him of the risk of death and injury. Yet which child wouldn't go off road into challenging terrain at the first opportunity? Which parents prohibit it?

A half-century elapsed from the rise of the automobile as an icon of American ingenuity, prosperity, and mobility until the mass production of recreational machines as new icons of technological verve, wealth, and off-road mobility. Similar processes of innovations in engine design, suspension, steering, and so on, changes in consumer buying patterns, and other factors enabled this mass production. By 1970 Americans had firmly established the phenomenon of motorized recreation. The small-bore two-stroke engine was the foundation of this recreation.

Recreation and Small-Bore Engines

Since the 1890s, scores of inventors have applied small engines to automobile, farming, marine, motorcycle, and other uses. Before World War II, outside of trucks and automobiles, which served both as pleasure craft and as utility vehicles, the major uses for internal combustion engines were for marine craft and farm equipment, not for recreational purposes.[2] Forestry, road construction, railroad locomotives, airplanes, and other applications were significant, but such uses took off after World War II.[3] Most of these vehicles and conveyances used four-stroke engines, while two-stroke engines were used more frequently on the farm and at sea in outboard motors and ultimately evolved into the small, reliable power engines used on recreational machines.

Gas tractors appeared at the turn of the century. They were bulky, cumbersome, and heavy but met a growing demand, particularly in wheat-growing sections of the country. They usually consisted of a large one-cylinder gas engine mounted on a heavy frame placed on four wheels. In the first decade of the twentieth century, designers sought lighter-weight gas tractors. The tractors came in an amazing array of shapes and sizes for use on farms of all kinds. World War I gave great impetus to the tractor industry because of increased agricultural demand and growing labor shortages as the war took able manpower abroad. By the eve of the Great Depression there were a hundred different companies marketing some 250 models and types of machines. According to the Census Bureau and other sources, there were 920,000 tractors in the United States in 1930, 1.6 million in 1940, and 2.4 million in 1945. The machines were both two- and four-stroke engines with one to two cylinders. A large number of the tractors were Fordsons.[4] Ford produced 1,227,694 of these in all, three-quarters of a million units of the Model F from 1917 to 1928 alone, more than any other tractor before or since. Recreational machine operators today are accustomed to electric starters and engines that are easy to start even in cold weather. But like the early recreational machines, the early Fordson tractor often required brute strength, in this case hand-cranking, to get going, especially when cold. Some farmers would build a fire under the tractor to warm up the crankcase and gear boxes to make it crank easier. Engaging the clutch (and listening to the gears grate into place) was another joy of the Fordson.[5]

Small recreational two- and four-stroke engines were also used in outboard motors. In 1907 Ole Evinrude, whose name still carries weight among marine motor manufacturers, built his first outboard motor. The design consisted of a horizontal cylinder, a vertical crankshaft, and a driveshaft with direction-changing gears housed in a submerged lower unit. The Waterman Porto Motor, which could be attached to a rowboat, also appeared about this time. According to the Evinrude Company, the Porto "was a dismally inferior product," and "the most enticing statement the manufacturers could think of to advertise it was 'Don't be afraid of it!'" Over the next two decades Evinrude produced a series of smaller, more powerful, and lighter-weight motors. A new two-cylinder motor, the Elto (Evinrude Light Twin Outboard), boasted 3 hp as compared with 2 hp for the one-cylinder Evinrude. It weighed only 46 pounds, 27 pounds less than the Evinrude, and substituted aluminum where possible for brass and iron. The company expanded following a successful national advertising campaign and rapidly growing overseas sales to Danish and Norwegian fishermen.

The Evinrude Motor Company was sold to Briggs and Stratton in Milwaukee. But in 1926 the Briggs and Stratton directors voted to leave the outboard motor field. Briggs decided to remain in that business and in 1929 joined Ole Evinrude to establish the Outboard Motors Corporation. (Lawnboy power mowers were also manufactured by Evinrude.) While Evinrude stressed "lightness of motor, ease of starting, smooth performance, and general dependability," he suddenly faced significant competition from a heavy but powerful engine produced by the Johnson Motor Company of South Bend, Indiana, that enabled boats to reach 16 mph while other motors could do no more than 10. As in the 1990s with the fascination with weight in SUVs, so in the roaring 1920s the fascination with speed, weight, and power made the Johnson motor a significant competitor. The depression led to a significant drop in sales, however, that persisted into the mid-1930s, and smaller, lighter, cheaper engines regained their popularity. The Johnson Company failed and was folded into the Outboard, Marine, and Manufacturing Company, which now manufactures Evinrude, Elto, and Johnson motors and accounts for about 60 percent of all motors sold.[6] Bombardier today owns Evinrude.

In the late 1930s and early 1940s Carl Kiekhaefer improved significantly on the reliability of earlier outboard motors by using a rubber water pump rotor that tolerated sand, silt, and vegetation; housing that protected drive shaft, waterline, and exhaust from exposure to the elements; and so on. Kiekhaefer displayed his "Mercury" engines for the first time at the 1940 New York Boat Show, which generated orders for over sixteen thousand motors. The war effort temporarily interrupted plans for expanding outboard motor production, but Kiekhaefer secured military contracts for air-cooled two-man chain saws for the army. In the postwar years many Americans translated the rise in their leisure time and disposable income into such activities as boating and water-skiing. Boating was no longer solely for fishermen. Mercury quickly resumed production to meet the increasing demand for larger and more powerful outboard motors. The company grew rapidly in the 1960s and remains a leading producer of outboard engines, with over six thousand employees.[7]

Yamaha and Honda developed outboard boat engines in the 1960s, around the same time they started to manufacture motorbikes and motorcycles. Yamaha's first outboard was the P-7, whose development commenced in 1958 and grew out of motor scooter and motorcycle engine development. Hence the P-7 would rely on components from the motorcycle division where possible; it would have an air-cooled engine and an adjustable drive unit length to make it mountable on all types of boats; and it was to be adaptable to burn kerosene as well as gasoline. The R&D team tested different materials for components and block, eventually settling on a high-silicon-content aluminum alloy that was lightweight and had good resistance to both abrasion and seawater corrosion. Problems in propeller design and adjustable drive also arose. The team eventually settled on a long version and a short version. Honda followed with a four-stroke outboard in 1964.[8]

For the most part two-stroke internal combustion engines have powered these utility or recreational applications and their postwar incarnations as gardening equipment, chain saws, snowmobiles, ATVs, and personal watercraft. Pressure to produce quieter and less polluting engines has led to their steady replacement by four-stroke engines or fuel-injected two-stroke engines. The two-stroke engine has such significant advantages as its relatively small size, lower cost, and mechanical simplicity. The two-stroke engine has lighter weight per unit of horsepower than a four-stroke engine and lacks the complex system of intake and exhaust valves, cams, and associated valve motions present in a four-stroke engine. In a two-stroke engine, an air-fuel mixture is drawn from the carburetor or fuel-injection system through a port into the crankcase. When the piston is forced down, the exhaust port is uncovered first, and hot exhaust gases begin to leave the cylinder. As the piston moves down, the intake port into the cylinder is uncovered, and the pressurized air-fuel mixture enters the combustion chamber. At some point both the intake and exhaust ports are open, which means the timing and airflow dynamics are critical to proper operation. Outflowing of exhaust gases create suction, and when the piston moves up higher and higher it finally closes off both ports, sealing the new fuel inside and compressing it. When it reaches roughly top dead center, the spark plug ignites the compressed fuel mixture, which burns and forces the piston down in the cylinder. The engine is lubricated by oil in the gasoline.[9] While the result is power with every downward stroke, the engine also expels significant quantities of unburned fuel into the environment.

Two-stroke engines have the requirement of thorough "scavenging" of the cylinder of burned—or shall we say incompletely burned—gases. This is because engine operation depends on "not only the volume of the fresh mixture that can be taken in, but also the explosibility of the charge. Too great a remainder of such gases not only seriously decreases the capacity of the machine, but it may go so far as to prevent ignition altogether."[10] In the ideal scavenging process the fresh mixture would push out the residual gases without mixing or exchanging heat with them, and this process would continue until all the burned gases had been replaced with fresh mixture, at which point the flow would cease. In practice, the ports remain open until well after the cylinder is at bottom center, and the pressure of the port closing is seldom as high as the inlet pressure. As a result, "some portion of the fresh mixture is usually lost through the exhaust ports."[11] There really is no ideal condition.

Two-stroke engines produce significantly higher pollution than four-stroke engines for another reason: they require the mixing of oil and gasoline for lubricating purposes, which leads necessarily to the burning of oil along with the gasoline and, since it burns incompletely, the expulsion of the unburned mixture into the atmosphere.[12] Most two-stroke engines now have oil injection, so the need for premixing gasoline and oil has been eliminated. The advantages of four-stroke engines are less pollution, lower fuel consumption, and more complete fuel combustion. Most important, in a four-stroke engine any residual gases will mix with the fresh charge to be burned more fully.[13] But the disadvantages of two-stroke engines were easily passed along to the consumer so long as gasoline prices were low and manufacturers did not have to consider clean air and clean water laws. Also, as noted, advances in exhaust tuning have made contemporary two-stroke engines much more efficient by limiting the loss of fresh charge through the exhaust port.

In early texts about internal combustion engines, engineers had already embraced the view that some loss of fuel and oil was inevitable, but since fuel was cheap and pollution laws were weak or nonexistent, this consideration was not viewed as crucial. This view prevailed into the 1970s, when regulators in the U.S. Environmental Protection Agency (EPA) first required automobiles to adopt technologies of efficiency and pollution control, followed in 1996 by the establishment of similar standards to limit pollution in small-bore engines. Emissions in today's engines depend on specifications, model year, horsepower rating, load factor, system design, and speed, hours of use, and frequency of tuning.

Today's two-stroke motors remain filthier than four-stroke engines even when they use fuel injection. They emit a blue-gray smoke composed of toxic and smog-forming compounds. Some ATV, snowmobile, and jet engines can expel up to 30 percent of their oil and gasoline unburned into air and water, producing as much as 4,000 times as much carbon monoxide and 118 times as many smog-forming pollutants as modern automobiles on a per-mile basis. Ninety percent of the 34 tons of smog precursors currently emitted each day by off-road motorcycles and ATVs comes from two-stroke engines. Many older two-stroke marine engines have the added disadvantage of draining excessive fuel from the crankcase directly into the water, a process as disgusting and wasteful as it sounds—hence scavenging devices that recycle the lost fuel and reduce oil throughput. Yet they continue to expel significant quantities of known carcinogens like benzene, toluene, ethyl benzene, and xylene in the unburned fuel, as well as carbon monoxide, nitrous oxides, particulates, hydrocarbons, polycyclic aromatic hydrocarbons, and the additive MBTE in the exhaust.[14]

Manufacturers had made great progress in their ability to build cleaner, more efficient engines even before being prodded to do so by the EPA. For example, the technology of fuel injection for efficient mixing and burning of the fuel-air mixture had been in development for over a hundred years, with patents first appearing around World War I.[15] The systematic application of fuel injection in two- and four-stroke engines during the 1990s was therefore tardy and imposed no hardship on manufacturers, despite what some of them claimed. Other manufacturers disingenuously claimed that they lacked the technology or capital to switch to four-stroke engines, even though such engines were hardly cutting-edge technology.[16] Unfortunately, while manufacturers are easily meeting the 1996 EPA standard, it will be decades before older engines in older vehicles have been retired to the trash heap.

Small-bore engines continue to contribute disproportionately to the nation's air pollution in already congested and smog-filled areas. In California, ORV riders fill trails and cover dunes adjacent to or in urban areas that already suffer from poor air quality, for example, in Hungry Valley (with impact on Los Angeles), Pismo Beach (adjacent to San Luis Obispo), and Ocotillo Wells (which contributes to San Diego air pollution). In cooperation with industry, California officials developed emission control regulations for ORVs (and PWCs, lawn mowers, chain saws, ATVs, golf carts, and the like). The regulations required use of catalytic converters, fuel injection, and other technology but allowed old equipment to be used and replacement parts to be available for them.[17] The success of California laws indicates that regulation can have the desired effects and will not impose undue hardship on manufacturers but rather benefits all citizens.

Despite the EPA ruling that required recreational and gardening machine engines to meet cleaner air standards, and the ability of manufacturers to do so, it has not been smooth sailing to clear, clean air. The manufacturers have sought to delay new engine efficiency and pollution standards. They claim that such standards place a costly burden on them that might lead to the loss of jobs to China and elsewhere. For example, in 2003 the Briggs and Stratton Corporation, the nation's largest maker of outdoor lawn and garden equipment, asserted that new California Air Resources Board standards requiring catalytic converters on the equipment, aimed at reducing air pollution from lawn and garden equipment and other engines, would require the company to "ship some work overseas." The company claimed that the California standards would lead to the loss of twenty-two thousand jobs in twenty-four states. Senator Christopher Bond (R-MO), who represents the home state of the company, then inserted a provision in an appropriations bill "to kill the California standards." Further investigation by the nonprofit Clean Air Trust resulted in criticism of Bond's provision, both for blocking the pending California standards and for taking away states' rights to require cleaner nonroad engines.[18]

But Bond and others were not finished. In June 2005 a Senate spending panel "approved language delaying a long-awaited federal rule aimed at curbing air pollution from lawnmowers and other small-engine machines." The amendment instructed the Environmental Protection Agency to conduct a six-month study into whether installing catalytic converters to reduce air pollutants from outdoor equipment would pose a safety threat. While this amendment failed, opponents of safety and environmental measures for small-bore engines have sought additional studies as a tactic to stall action. In this case, Senator Bond asserted that the catalytic converters were a potential fire threat. But the EPA determined they simply are not, noting that "catalytic converters reduce harmful emissions by as much as 75 percent."[19]

In addition to concerns about pollution, some people including machine operators have also urged manufacturers to improve noise characteristics of the engines and to make available noise data so that consumers can make informed choices about purchases. Recreational machines tend to disrupt lives and ecosystems through their noisy engines. Modern-day life at work and at home is already so very noisy, especially in urban settings. Americans have long used their automobiles to escape that noise for the quiet of parks and wilderness areas. Fordist recreation means that play and rest outside are noisy as well. According to a variety of sources, if the noise around you forces you to raise your voice to make yourself heard a meter away, your hearing may be at risk. Of course, recreational machines require you to raise your voice to be heard, even when they are 50 meters away. Noise levels of normal conversation, measured in decibels (dB), range from roughly 60 dB to 65 dB. The decibel scale is logarithmic: 65 dB is the level at which you have to raise your voice, and prolonged exposure to noise above that level significantly damages hearing. Heavy traffic produces 80 dB, lawn-mowing 90 dB. Specialists argue that laboratory measurements of noise levels are somewhat misleading when compared with noise heard in real-life settings. The fact remains that if noise cannot be reduced or removed at its source, then personal protective hearing equipment (earmuffs, earplugs, headphones) should be worn.

Recreational machines produce noise levels of 85–100 dB. Manufacturers have tried to tone them down, with some success. New-model machines mostly reach laboratory standards of 96 dB—loud enough to produce loss of hearing and to suggest that ear protection always be employed. Manufacturers of personal watercraft have placed the exhaust under water to quiet them—a solution that has been only partly successful, since PWCs jump out of the water, generating a constant "whomp, gurgle, whoom, slap, whomp, gurgle, whoom."

Federal, state, and local public health authorities have long recognized the dangers and costs of exposure to excessive noise. But governmental activity in the area of noise abatement prior to the 1970s was almost nonexistent, and in any case was short lived. In Europe, policy makers continue to recognize the crucial need to address noise problems head on. The European democracies are far ahead of the United States both in recognizing the public health dangers of noise and in coming to grips with it. They understand that it is cheaper to abate than to deal with the health consequences.[20] In the United States, specialists and policy makers recognize that noise from power tools, highways, industrial processes, boom boxes, and such can be anything from a public nuisance to a health hazard. But beyond acknowledging that a problem exists, they seldom act forcefully.[21]

The passage of the National Environmental Policy Act (NEPA) in 1969 required agencies to consider noise as well as air and water pollution in environmental impact statements. Congress directed the EPA to establish an Office of Noise Abatement and Control (ONAC) to prepare recommendations for legislation. Congress acted despite the lack of any public demand for legislation for two reasons. First, railroads, interstate motor carriers, and motor vehicle manufacturers were concerned about complying with conflicting state and local regulations, and federal ones might resolve the conflicts. Second, EPA officials had reported to Congress that 34 million persons were exposed to nonoccupational noise capable of inducing hearing loss, 44 million felt the impact of aircraft and other transportation noise, and 21 million lived through construction noise.[22]

The EPA gained responsibility "to promulgate emissions standards, require product labeling, facilitate the development of low emission products, coordinate federal noise reduction programs, assist local and state abatement efforts, and promote noise education and research." While implementation was difficult, ONAC accomplished a great deal. In early 1981, however, its director learned that Reagan's White House had arbitrarily determined to end its funding, in an ineffective attempt to control burgeoning deficits by cutting environmental and social programs. Congress hoped to shift noise control to state and local governments—a doomed policy, given the need for federal standards and resources. Shockingly, of the twenty-eight environmental and health and safety statutes passed between 1958 and 1980, only the Noise Control Act of 1972 was fully stripped of funding.[23] By the time it was disbanded in 1981, ONAC had established without question the links between costly public health problems and excessive noise.[24] In this context many specialists question the assertion of recreational machine manufacturers that their machines have no public health costs or are adequately quiet because they meet standards they themselves have set. An August 1978 report provided evidence of hearing loss, heart disease, and high blood pressure, ulcers, and other illnesses, as well as low birth weight. Noise also contributed to crime via impacts on education and social development.[25] Unfortunately, unthinking deregulation won the day, with direct impact on the quality of recreation in America: recreational areas are filled with noisy machines. Today's recreational vehicles meet voluntary standards established by industry in consultation with the government. Many experts and nonusers consider those standards inadequate.

Without ONAC it has been difficult to regulate such new, persistent, and dangerous sources of noise as ATVs, personal watercraft, and snowmobiles. How have states filled the void left by the federal government? In 2003, California, frequently a leader in public health and safety regulation having to do with environmental issues, set strict noise regulation standards for off-road vehicles and ATVs operated in state vehicular recreation areas. The California Department of Parks and Recreation reduced noise emissions from a 101 dB standard to 96 dB,[26] still loud enough to cause loss of hearing unless operators use ear protection. The regulations also permitted noise levels up to 101 dB for off-road vehicles manufactured before 1986 and "competition" ORVs manufactured before 1988.[27] Since most of the public health impacts of noise had been identified by the end of the 1970s, noise abatement should not be considered a scientific problem but primarily a policy problem requiring action. Representatives of the Noise Pollution Clearing

House assert that "the air into which second-hand noise is emitted and on which it travels is a 'commons,' a public good. It belongs to no one person or group, but to everyone. People, businesses, and organizations, therefore, do not have unlimited rights to broadcast noise as they please, as if the effects of noise were limited only to their private property. On the contrary, they have an obligation to use the commons in ways that are compatible with or do not detract from other uses." Those who ignore or downplay the impact of their noise on others are "in many ways, acting like a bully in a school yard . . . they disregard the rights of others and claim for themselves rights that are not theirs."[28] Noise exposure is on the increase. Permitting it to spread throughout park and wilderness environments is not a solution. Nor is requiring hikers and recreational machine users alike to use ear protection (though indeed they should).

Public Health, the Environment, and Recreational Machines

Lawmakers have been equally slow to recognize other public health costs of recreational machine use: injury, trauma, and loss of life. Whether ATVs, snowmobiles, or personal watercraft, recreational machines have contributed to a crisis in public health. The crisis involves new kinds of injuries, usually involving severe trauma, internal bleeding, broken bones, disfigurement, maiming, and in many instances paraplegia and quadriplegia; growing numbers of deaths; and high costs associated with hospitalization, treatment, rehabilitation, reconstructive surgery, and lost hours of productive work. The American Academy of Orthopedic Surgeons estimated the total cost of injuries from ATV accidents in 2000 alone at $6.5 billion. Sadly, children have borne the brunt of this crisis. They rarely understand their own strengths and weaknesses, let alone how to operate a fast, heavy, versatile machine. Between 1982 and 2001 at least seventeen hundred people died in ATV accidents, eight hundred of them children under the age of twelve. Ninety-five percent of those children injured or killed were riding an adult-sized ATV. If adults, and if children whose parents aren't paying attention, wish to engage in risky behavior, there is little we can do to stop them. But is this simply an issue of liberty (the freedom to ride as one pleases) versus nonliberty (government intrusion into private lives)? The operation of recreational machines carries risks to our own safety and that of others. The costs in human suffering and medical care are great, the heartbreak to family and friends of individuals injured or killed substantial.

Why do Americans accept the risks inherent in operating recreational machines? People who have studied risk argue that the more familiar a risk is and the more voluntarily a person embraces it, the less likely he or she will worry about the danger. Hence, Americans gladly hop into their automobiles, many of them without using seatbelts, knowing that 100 to 150 of them will die in accidents every day. Yet they insist that such unfamiliar technologies as chemical additives to ensure a safe food supply should be restricted on the grounds that a handful of people may be sickened or injured by that additive. They gnash their teeth over small—sometimes infinitesimally small—and novel risks. In addition, many individuals equate regulation of recreational machines, or requirements that operators follow safety rules, wear a helmet, and otherwise use common sense, with deprivation of liberty. Like the automobile, ATVs are familiar, their use is voluntary, and they symbolize freedom. On top of that, they were already well embedded in our social, political, and cultural institutions by the time state and federal agencies had analyzed anecdotal evidence and gathered sufficient data to reveal the significant risk of operating them, especially three-wheeled models.

Fordism in Nature

The spread of recreational vehicles throughout the nation's lands with what many individuals argue is insufficient oversight has created one last situation that calls for circumspection. What has been the impact of recreational vehicles on various ecosystems? Ought we to prevent machines from entering places they haven't yet entered? Should we roll back permissions where they have contributed to destruction of wolf, elk, loon, and wading bird habitat?

What are the costs of regulating the access of recreational machines to parks, wilderness areas, and other settings? How do we balance the rights of operators and nonoperators?

Recreational machines have both direct and indirect environmental effects. Irresponsible operators use them to harass animals and destroy property. Even proper use frequently results in significant environmental damage. The machines are mobile and versatile, extending the impact of recreation far beyond past limits. They produce ruts and mud, destroy root systems, and accelerate erosion in all environments. Snowmobilers, while operating on a foundation of several inches or even feet of snow, contribute to degradation by frightening wolves, elk, and deer and by damaging seedlings and root systems when they are at their weakest. Snowmobiles, ATVs, and jet skis collectively go everywhere, at any season, on snow, land, and water. They open previously inaccessible wilderness areas and nature preserves to more and more people.

What is the relationship between machine users and nonusers? As a *Yale Law Journal* article indicated as early as 1973, hikers and other low-impact visitors to natural sites may enjoy an area "blissfully unaware of each other's presence, [while] a single noisy snowmobile can disrupt the enjoyment of all of them, though its operator may be well-intentioned and indeed may not even know that others are in the area." The authors criticize the ad hoc way in which machine access to nature has evolved. While focusing on snowmobiles, their conclusions hold for ATVs and jet skis as well: "The system . . . fails to insure that the costs of snowmobiling are paid by those who cause them. Snowmobiling . . . in effect receives a subsidy from those whom it causes damage and annoyance. It consumes great quantities of scarce goods like quietness, natural resources and recreational opportunities."[29] Wilderness disappears, and outdoor activities become less desirable.

In the remainder of this book I examine the impact of recreational machines in a variety of ecosystems and settings. But it should be pointed out that the appearance of recreational machines in parks dates to the founding of the national park system in the early twentieth century, when officials ordered the construction of roads and visitor centers to facilitate automobile traffic. Machines have long been a part of recreation in the United States, as have federal efforts to facilitate machine access. This may help explain why conflicts over access of recreational machines have been so difficult to resolve: users and regulators have good reason to think that it is normal for them to be in parks. Several critics have worried about this "normal" state of affairs from the start. Bob Marshall, a hiker and explorer extraordinaire with two forestry degrees and a Ph.D. in plant physiology, founded the Wilderness Society in the 1930s while he was an employee of the Department of the Interior. He immediately admitted the negative impact of machines on nature, especially on federal lands that were supposed to be preserved for a variety of uses. He was disturbed to find that everywhere he looked, wilderness had been cut back. In 1930, in *Scientific Monthly,* he called for "an organization of spirited people who will fight for the freedom of the wilderness." Marshall worried that the government had failed to protect natural resources. He believed that the conservationist ethos—that of ensuring the fair use of natural resources among many different parties and interests—had been overtaken by a development ethos.[30]

The National Park Service stood with the Forestry Service in securing access of machines to federal lands. In the first three decades of the twentieth century, and especially under the first director of the National Park Service, Stephen Mather, the service built tunnels in giant sequoias, dumped soap into geysers to make them erupt, fed garbage to bears to half-tame them, and built roads and visitor centers in order to attract tourists. In the 1930s, as part of the effort to get people back to work through government programs during the Great Depression, the Forest Service, Park Service, and other federal organizations built highways: Skyline Drive through Shenandoah National Park, a parkway in the Great Smoky Mountains National Park, a highway connecting them, and so on. Marshall attacked Skyline Drive as a "gigantic, artificial parking place" that exterminated the wild mountain meadows in New Found Gap. Instead of wilderness, he found scores of automobiles, many of them blaring "jazz on the radio as a substitute for the primitive." In New York, the state used Civilian Conservation Corps funds to build truck trails in Marshall's beloved Adirondacks that opened the way for today's ineluctable spread of ATVs in once beautiful parks. New York's Department of Conservation intended to build 120 miles of these trails. Were they needed for fire protection? Or did roads simply bring automobiles and motorists who posed a fire danger with their campfires and cigarettes? Marshall properly wondered. Primitive areas disappeared "with appalling rapidity. Scarcely a month passes in which some highway does not invade an area," he wrote. Several areas were later closed to traffic. But once the roads were built, the machines congregated.[31] Snowmobiles, ATVs, and personal watercraft, millions of them, have now been added to the mix, and federal officials today work to facilitate their access, not restrict it.

Marshall provided one of the earliest and most prescient discussions of the nature of wilderness and why it must be preserved against the encroachment of such hallmarks of civilization as roads and machines. Marshall defined wilderness as "a region which contains no permanent inhabitants, possesses no possibility of conveyance by any mechanical means." This means that "all roads, power transportation and settlements were barred, but trails and temporary shelters, which were common long before the advent of the white race, are entirely permissible." Marshall understood that a philosophy of progress provided the germ for unabated disruption of nature, and that the more "progressive" a society, the more it had altered nature. Marshall called for a deliberate policy to preserve wilderness. The goal was to balance the happiness that will result if a few "undesecrated" areas are saved against "that which will prevail if they are destroyed."[32]

Marshall identified three benefits to society of preserving wilderness that remain pertinent today: the physical, the mental, and the aesthetic. By physical, Marshall meant the health benefits of access to pure air and quiet surroundings as well as the self-sufficiency that wilderness promotes. By mental, Marshall meant the "incentive to independent cogitation," the way in which wilderness promotes the repose, reflection, and relaxation that enabled such "virile" minds as Henry David Thoreau and John Muir to retreat into thought. (Dare we add Rachel Carson to the list?) And by aesthetic, Marshall meant the physical beauty of nature. He recognized that the physical beauty of nature is a somewhat subjective consideration but nevertheless urged us to consider the "stupendousness," immensity, silence, and nontemporal constancy of nature. He wrote, "Any one who has stood upon a lofty summit and gazed over an inchoate tangle of deep canyons and cragged mountains, of sunlit lakelets and black expanses of forest, has become aware of a certain giddy sensation that there are no distances, no measures, simply unrelated matter rising and falling without any analogy to the banal geometry of breadth, thickness and height."[33] The government had long provided funding for museums, galleries, concerts, gardens, parks, golf links, and menageries, expenditures that satisfied "only a fragment of the community." But these programs were almost universally approved, Marshall maintained, and so would appropriations for wilderness areas. Indeed, surveys have consistently found American support for wilderness preservation and environmentally sound policies to be at the highest level.[34]

Unfortunately, lawmakers have ignored Marshall's concerns. They have seen land as inexhaustible. In the face of pressures for multiple use, they have rarely given adequate consideration to the importance of preservation. This means they have tried to balance lumbering, hunting, hiking, recreational vehicles, maintenance of wilderness, and other purposes but have not comprehended how one activity might interfere with others, and that several of them, road building for instance, degraded those areas for everyone. Even more, lawmakers have often treated the management of federal lands in the same way they have treated such health and environmental problems as noise abatement. They instruct officials to measure the impact of ATVs, snowmobiles, and jet skis and to come up with regulations to control that impact, yet they provide miserly funding for data collection and enforcement, leaving those federal lands in recognizably worse condition and their managers powerless to do anything about it. The fact that other machines, processes, and uses have negative effects does not lessen or justify the destruction wrought by recreational machines. Habitat has been altered or destroyed, species have been threatened, biodiversity has declined. The federal government seems unable to act, and in its abdication of responsibility, regulation has devolved to the states (which are often too overburdened and impoverished to manage access of recreational machines to public and private lands) and to the industries (which have often co-opted the regulatory agencies and processes). In the face of voluntary limits and regulations, tens of thousands of miles of legal and illegal trails have been pushed through forests and over dunes, hundreds of thousands of acres of land have been irreparably damaged, as wilderness has been taken over by machines and their purveyors.

The federal government manages 629 million acres—over 900,000 square miles, or more than one-quarter of the nation's land, the rest being in state, corporate, and private hands. The Bureau of Land Management handles 264 million acres; the Forest Service (of the Department of Agriculture), 193 million acres; the Fish and Wildlife Service, 93 million acres; and the National Park Service (of the Department of the Interior), 80 million acres. Generally the first two agencies give carte blanche to recreational vehicles, while the latter two are much more restrictive, and no wilderness area or national monument permits the use of off-road vehicles.

The determination of the Bureau of Land Management and the Forest Service to permit use of machines on federal lands grew out of Progressive Era (1890–1914) notions of managing lands, and the forests, water, and other resources on them, so as to achieve the greatest good for the greatest number of people. That is, land managers sought to ensure multiple uses across competing demands—resource management, resource development, and recreation. They used scientific and engineering techniques to do so, studying how much of a particular resource was available, what the demand for it was, and how to distribute it among all users over time. They overestimated the ability of objective science to adjudicate those competing demands. Like many of their counterparts today, they failed to understand that the determination of access to resources must necessarily be a political decision, for there will inevitably be winners and losers of access. And while their goal—a science-based conservation of resources and access to them for present and future generations—was a noble one, they erred on the side of access and resource development. Congress facilitated this error by offering federal lands to miners, cattle raisers, and others at subsidized rates. Further, land managers rarely calculated how one use—say, the construction of roads to permit logging—might have a major and irreversible impact on the other uses.

Protection of the environment does not fit easily into the framework of determining what is the greatest good for the greatest number of people. Environmental concerns have often gotten short shrift in the face of other pressures. Many, many more citizens wish to experience nature than to preserve it, and most of them use machines to do so. Add to this the fact that recreational pressures have grown manifold with increases in population, disposable income, and numbers of machines. No wonder that federal, state, and local officials face difficult decisions about what to do.

The "Tread Lightly!" campaign typifies the difficulty of managing conflicts involving machine access to parks and recreational areas. The campaign may be an example of agency capture, a concept I define more broadly than the academic literature as what occurs when business and other entities come to dominate the agencies established to regulate their activities in the name of the public good and then use those agencies to promote regulations that serve their own private economic interests. Public health advisories and warnings come in the form of small print, illegible warning stickers, and other devices that do more to facilitate marketing and lessen liability than to regulate. While some agencies remain true to their statutory missions, others begin to substitute promotional activities for regulatory ones or lack the vigilance or resources to carry out their missions.

An initially promising and ultimately faulty approach to off-road recreation with powerful machines, Tread Lightly originated in the Forest Service in 1985 in partnership with manufacturers and environmental groups. Each group hoped to avoid conflict over questions of access to public lands by guaranteeing some lands in exchange for responsible use. What was "responsible use"? Tread Lightly involved a five-step path to recreational awareness:

Travel and recreate with minimum impact
Respect the environment and the rights of others
Educate yourself, plan and prepare before you go
Allow for future use of the outdoors, leave it better than you found it
Discover the rewards of responsible recreation

Forest Service administrators believed that they could avoid difficult decisions and litigation if environmentalists accepted this educational program without argument. Manufacturers sought access to public lands for the skyrocketing number of machines they produced. And environmentalists believed that an agreement to produce ethical advertisements promoting responsible machine recreation would help to limit degradation of sensitive areas and discourage irresponsible operators from illegally entering closed areas.

At an early stage, mainstream environmentalists persuaded manufacturers not to advertise using photos or artists' renditions of happy recreationists ripping through mud or over sand dunes. Dozens of ATV manufacturers (Honda, Suzuki, Bushwacker, Polaris, Yamaha) and all of the major automobile manufacturers signed on. The guidelines served as a reference for creating advertisements that were "powerful, yet not destructive."[35] Manufacturers were supposed to show operators riding in legal areas, staying on trails, never blazing new ones, heading for an obvious destination, avoiding water and wet trails, riding quietly and at prudent speeds so as to leave no trace.[36]

The Bureau of Land Management took over the program from the Forest Service before the almost inevitable "privatization" of Tread Lightly occurred. Since the early 1980s under President Reagan an anti–big government movement has attempted to allow market mechanisms to provide services more efficiently, if not more justly, and to regulate the safety of technologies based on the assumption that consumers control production through demand and by refusing to purchase unsafe or annoying products. In 1990 the government "transferred" Tread Lightly to the private sector, where it continues to operate as a not-for-profit organization. In 1997 it assumed responsibility for water-based recreational activities. The U.S. Army Corps of Engineers, National Park Service, and Fish and Wildlife Service continue as partners in the organization. They have joined state agencies, media, clubs, and individuals who wish "to protect the great outdoors through education." Two problems persist. First, the organization is dominated by manufacturers. Second, there is no way modern recreational vehicles can travel with minimum impact, as today's advertisements indicate they can.

Advertisements for ATVs and other such vehicles show young (usually white, middle-class) families, with healthy, muscular bodies and big smiles on their faces, mounted on muscular vehicles charging through mud, water, or snow. The advertisements equate responsible machine recreation with high-speed exploration of complex ecosystems. Through unethical and misleading messages, they depict machine as victor and nature as enemy. Brochures from Kawasaki, Honda, Polaris, and Yamaha come with the Tread Lightly sticker but clearly do not encourage such behavior. The "action photography" involves "professional riders" on closed courses. Consumers assume they can operate their machines at the speeds and in the wilderness places that professionals do. In one set of brochures the Kawasaki KFX700 creates "the perfect sand storm." The KFX400 gives "maximum exhilaration." Ads for Kawasaki's 2005 "sport utility" ATVs proclaim that "nothing beats the use of brute force," showing a machine on rocky cliffs above a beautiful mountain lake in the middle of deep forest. "The hunt for the king of ATVs is over," claims another ad showing a camouflaged ATV enabling a hunter to bag game in a previously inaccessible area in the Rockies. And an operator moving through deep mud demonstrates "Big power. Bigger authority." The ATV operator controls nature. According to the Izaak Walton League of America, which since 1922 has fought to protect the rights of all

Americans against unethical hunters and anglers who disgrace the honorable, responsible use and enjoyment of the nation's resources, Tread Lightly "has served some manufacturers as little more than useful cover and stamp of approval for their increasingly disturbing messages." One Izaak Walton study demonstrated the ubiquity of unethical advertising inconsistent with Tread Lightly guidelines.[37]

In the case of Tread Lightly the agency capture involves "logo hijacking." Consider the publications concerning four-wheeling "lightly." The professional production and matter-of-fact language of these publications mask the costs and risks of off-road four-wheeling. In one pamphlet, a section on "negotiating terrain" warns drivers to drive slowly, to avoid sudden acceleration, turns, or braking, and to head straight up or down any hill or grade. The accompanying photo shows a HUM-V at a 38-degree angle. The brochure also instructs drivers to avoid obstacles, rocks, and mud, and not to spin tires or gun engines. But as the advertising brochures indicate mud is what it's really all about. Tread Lightly in fact encourages riders to cross streams and drive through ruts, gullies, and trail washouts.[38]

Tread Lightly offers absurdist "minimum impact" tips for ATV and snowmobile off-roading. According to one pamphlet, "ATV riding is a wonderful way to see the outdoors and, if done properly, an environmentally sound way to experience the backcountry." Further, "Anyone can ride fast but it's the skilled rider who can ride slowly over challenging terrain with minimal impact to the ground." The keys are to avoid wet or muddy conditions. Use low throttle. Regarding streams, cross only at fording points, and don't blast through, because it's bad for fish and other aquatic life. The pamphlet claims that switchbacks—sharp zigzags that minimize grades and so facilitate access—prevent a trail from becoming a "miniature river" during a rainstorm. Yet switchbacks also lead to erosion and to the destruction of more surface area and therefore more land. Instructions to the operator not to make ruts deeper and to avoid meadows and marshy areas indicate there can be no "minimum-impact" ATV use.[39]

Regarding snowmobiles, manufacturers on the one hand encourage such wholesome fun as camping, ice fishing, photography, and organized club activities. Yet they also urge riders to seek access to "hidden woods and distant mountains."[40] Such exploration leads riders into encounters with megafauna. Tread Lightly philosophers inform us, "Some animals, especially large, heavy ones such as buffalo and moose, use groomed, packed trails as handy walkways. If you see them on the trail, remain a safe distance away and they will eventually move off the trail and let you pass."[41] Anecdotal evidence indicates that operators often, almost unavoidably, frighten animals by gunning engines and so force them to expend precious energy on flight. Scientific data discussed in subsequent chapters reveal harm to the creatures.

A Forest Service publication for hunters and their children reveals the disjunction between ATV and other uses of the environment and the Orwellian language of Tread Lightly. The handbook tells the story of a father instructing his son about how best to hunt his first deer. The boy and father hunt safely, carefully, approaching the buck quietly. Just as the boy raises his rifle to kill the buck, an ATVer accelerates through a nearby field, frightening the animal away. Tears well up in the boy's eyes, and the father has a hard time explaining why such discourteous recreationists have become the norm.[42]

In February 1972 President Richard M. Nixon, whose presidency was in grave difficulty owing to the Watergate scandal, found the time to build on his already substantial environmental record. He had signed the National Environmental Protection Act in 1969, which led to the establishment of the EPA. Now he issued Executive Order 11644, "On Use of Off-Road Vehicles on Public Lands," to control and direct their use so as to protect resources, promote safety, and minimize conflicts among users. The order stipulated locating trails and recreational areas in such a way as to minimize the machines' damage to soil, watersheds, vegetation, and wildlife habitat, and it prohibited their use entirely in wilderness or primitive areas.[43] In 1977 President Jimmy Carter signed Executive Order 11989 to follow up on Nixon's mandate. That order directed land managers to close land to off-road vehicles where their use "will cause or is causing considerable adverse effects on soil, vegetation, wildlife, wildlife habitat or cultural or historic resources of particular areas or trails of the public lands," until the adverse effects have been eliminated and measures taken to prevent future reoccurrence. Conflict over these executive orders continues into the twenty-first century.

In the following three chapters I explore the spread of recreational machines into the fabric of postwar American life and leisure. I consider the history of the recreational machines and their clubs and discuss their associated safety and environmental issues. I consider efforts of responsible operators to police their activities through clubs and in cooperation with local and state officials. Such individuals have helped to develop a regulatory, licensing, and registration system to encourage safe and appropriate use and to establish safe and appropriate trails. Yet I worry that the machines unavoidably have impacts on the environment that call for greater regulation. Their noise, their pollution, their inherent instability at the speeds and in the terrain for which they are intended, have made them a growing problem that calls for a national discussion. Today's recreationists seek out nature on high-speed vehicles that can go far off the beaten track, deep into the pine forests of northern Michigan, the grassy plains of Minnesota, the tidal basins and estuaries of the Florida peninsula, the lakes of New England, the sand dunes of southern California, the arid Redrock Wilderness of Utah, and the national forests, parks, and wilderness areas. With each ATV pass, operators see the remnants of Native American culture; with each snowmobile pass, the remnants of Inuit and Saami lifestyles. Snowmobiles compact snow and the soil underneath, changing the ecosystem. Jet skis threaten shallow waters in lakes and estuaries, with significant impact on loons, other wading birds, and their nests. ATVs destroy streams and creeks and the fish in them. With every pass through the woods, the paths and trails become muddier, wider, deeper, less able to recover. On these fast-moving recreational machines powered by small-bore engines, how much of nature do their operators see? Do they recognize the damage to the environment their machines produce? Is that damage at an acceptable level? It is no surprise that Americans—those very people who embraced the automobile and the assembly line first and better than all other people of the world—have embraced the recreational machine. But instead of nature or wilderness, they now find a mechanized, motorized landscape. They have accelerated the disappearance of the very vistas, frontiers, parks, and wilderness areas they set out to see. They do this with every pass, in increasing numbers and on increasingly powerful machines.

‹ 2 ›

ELK-SNOWMOBILE SYNDROME

FLAT IN EVERY DIRECTION, except for gullies, streams, and rivers, and for grasses and trees planted as shelter belts along property lines, the upper midwestern states offer a perfect environment for the snowmobile. When the ice receded ten thousand years ago, it left behind rich soil in what had once been Lake Agassiz. What is now northern Minnesota and North Dakota was settled by Norwegian and Swedish farmers. An 1895 federal study indicated that agriculture would always be "the chief industry and source of wealth throughout the prairie portion of the area of Lake Agassiz." The fertile land benefited both the "farmer and herdsman," as it had "its former herds of elk, antelopes, and buffaloes." It supported "rapid progress in the production of wheat and other crops and in stock raising and dairying." Water resources including the Red River and thousands of lakes could be tapped year round.[1] To this day most people in the area still make their living through agriculture. It was hard work to bust the sod, and the growing season was short. Until recently, bitter cold and wind kept families inside for much of the winter. Residents still joke that they have nine months of winter and three months of sledding, but today the snowmobile has extended the sledding to six months. The snowmobile contributes to the fabric of everyday life. It's a utility vehicle, a way to get out to feed the animals or to meet friends, to get to school and work or to relax, or simply to avoid feeling shut in during the long winters. No wonder the Polaris slogan is "The way out."

In the small agricultural communities that dot the plains, ATV and snowmobile owners stress the utilitarian purposes of their machines. Those purposes reflect ways of life that have evolved over decades and the requirements imposed on residents by geography and climate. In much of the region the snowmobile will often be the most reliable (if not the only) form of transport during the bitterly cold winter months. Temperatures fall to −40°, and strong winds create deep and treacherous drifts. Snows average 60 inches a season, somewhat less than the 80 inches in much of New England, but more than enough for modern snowmobiles. When the cold days sock people into their modest houses and cabin fever sets in, the snowmobile enables neighbor to see neighbor. On the weekends, club members gather to hold races and rallies and to mark, repair, and build trails. Farmers use snowmobiles to haul hay to their cattle, or they use ATVs that can also work as snowplows with the addition of a $600 blade attachment.[2]

Snowmobiles skyrocketed in popularity in the late 1960s and 1970s. Their numbers increased from twenty thousand to several million in a decade. Their attraction was as recreational vehicles, although their inventors and promoters saw them for such utilitarian purposes as law enforcement and medical emergency evacuation, logging, communication, and other activities. Snowmobiles indeed facilitate emergency response and promote business and tourism in harsh continental climates. Hotels, restaurants, and gas stations might fold during the winter without them.

Snowmobiles have also contributed to an ongoing dispute about what wilderness is, for whom it exists, and what role state and local governments should have in protecting it. Whenever new machines arise that enable people to ride into previously inaccessible areas, they trigger both enthusiasm from Americans with the disposable income to buy them and who transform transportation through wilderness into a nearly unbounded recreational activity, and angry opposition from others who resent the noise and pollution that accompanies machine recreation. Snowmobiles may be the least offensive of the recreational machines, but their undoubted environmental impact and growing public health costs suggest that all is not well. The joy that snowmobilers rightfully take in their classic 'Cats and Ski-Doos often distracts attention from the deleterious effects on ecosystems and wildlife.

Snowmobiles provide a prime example of machines whose initial purpose as a utility vehicle has been almost forgotten in its newfound use as a toy. Yet this is no simple toy but a bullet-like marvel whose top speeds of 90 mph present challenges even to competent drivers. Snowmobiles are responsible for scores of deaths and tens of thousands of injuries annually. When manufacturers introduced them, there were few controls over their safety, noise, and environmental performance. They were spartan, not very comfortable, had perfunctory lights and hard steering, lacked sophisticated suspension, and often had no windshield. For years people used them without any thought to helmets, special gloves, or boots. Industry and snowmobile clubs encourage safe operation, proper equipment, and educational programs. Of course, manufacturers intend to produce vehicles that are safe, because if they are not safe consumers will cease to buy them. But they will always be thrillingly fast to operate and therefore dangerous. The Consumer Product Safety Commission estimates that each year about 110 people die while riding snowmobiles and that snowmobile accidents account for about 13,400 hospital emergency room–treated injuries. Approximately two-fifths of the reported deaths result from collisions with trees, wires, bridges, and other vehicles. Some deaths occur when the snowmobile rolls to the side in a ditch or stream and pins the operator under the vehicle. Others occur when the snowmobile enters water, often when the operator is driving across thin ice.[3]

As for environmental issues, state and federal officials have been slow to respond. Snowmobiles appeared in small numbers out of small fabrication facilities in the 1950s (although their predecessors date to the 1920s). By the end of the 1960s they were being produced in sufficient numbers to raise concern about their environmental costs in wilderness lost, magnificent megafauna disturbed, and forest rangers exposed to noxious emissions. Manufacturers and clubs have successfully fought most attempts to restrict snowmobile use, even gaining state funding to pay for their trails. Yet researchers have long known that snowmobiles compact snow, that compacted snow thaws later in the spring, and that compaction of the soil underneath pushes moisture and oxygen out of it. Studies of experimental plots and of trails and fields planted with timothy, clover, and alfalfa indicate significantly lower yields of these forage crops after winter snowmobile use. These were not anomalous results due to fluctuations in snowfall. Field studies from Maine to Minnesota and Manitoba show damage by snowmobiles to hardwood saplings and pines, even with a single pass. Deer and their predators that frequent snowmobile areas lose their natural wariness and orientation, reacting violently to the snowmobile invasion, likely expending energy and "upsetting the delicate balance between stored body fat and harsh winter environment." A single snowmobile has a far greater effect on ecosystems, flora, and fauna than fifty people on snowshoes. Snow cover, important to the survival of small animals (voles, shrews, mice, squirrels, skunks) by protecting them from exposure and predation, is disturbed, leading to nearly 100 percent mortality rates.[4] Agriculture has suffered as well; blueberry farmers in eastern Maine reported damage to their lands by the early 1970s.[5]

In the late 1990s researchers at Montana State and Michigan State universities poked around in elk and wolf skat. Immunoassays of fecal glucocorticoid showed that levels of the hormone were higher in wolves and elk after exposure to snowmobiles, indicating that the animals felt stress. Hormone levels were higher during snowmobile season, and levels rose and fell with daily snowmobile traffic.[6] While the researchers were unable to say conclusively that snowmobile activities affected population dynamics, the evidence seems to indicate that animals, snowmobilers, and other outdoor enthusiasts alike will suffer if we wait any longer to restrict snowmobile access to some federal and state lands.[7]

Ultimately, snowmobile manufacturers have managed to persuade short-sighted allies in Congress and the White House to postpone many regulations. Those allies embrace the time-tested expedient of postponing action by calling for yet more study of well-established phenomena (such as the adverse impacts of global warming). The goal is to keep national parks, in particular Yellowstone National Park, open to snowmobile operators so as to benefit local businesses at the entrances to the park, other recreationists be damned. Park rangers now have to wear respirators, hikers have lost any sense that they are in America's national forests, and elk and wolves must deal with industrial stress.

Snowmobile History

Early Power Sleds and Snowboggans

The ancestors of snowmobiles appeared in the late nineteenth century in the form of power sleds. They used half-track designs and often mimicked existing rail (locomotive) and road (automobile) vehicles. Antique-snowmobile historian Steve Pierce points out that the U.S. Patent Office awarded a certificate for a power sled to the Runnue Brothers of Crested Butte, Colorado, in March 1896. The sled had an "endless track of chain and eight steel crossbars supported by spring straps." The cross blades had spurs on the outer edges to get a good grip on ice. Another distant relative of the snowmobile was the Lombard steam hauler, perhaps the first tracked crawler vehicle to receive a patent. It was a kind of steam locomotive on treads that enabled lumbermen to move timber from the forest to the processing yard. Its inventor, Alvin Lombard of Waterville, Maine, followed with gasoline-engined versions of these vehicles, some of which made it as far as Arkhangel'sk province of the Tsarist empire.[8] Several early inventors also attempted to develop air-propelled snow vehicles, but they did not fare well against tracked vehicles.[9]

Between 1896 and 1930 several dozen inventors produced self-propelled (engine-powered) sleds. In 1904 W. H. Anderson of New York received a patent for a runner attachment for automobiles. He incorporated an endless chain fitted with lugs or spurs running inside the rear runners and powered by the automobile's driving axle. In 1912 R. Carroll of Simmonsville, Rhode Island, followed with a similar vehicle. Rudolf Kubelka introduced a "sleigh attachment for vehicles" that involved runners fitted under the front wheels and inside the back wheels that had a traction chain running around them like a half-track. In the 1910s Ernest Splittstoser of Pine City, Minnesota, mounted a Model T body on a wooden farm sled. He attached a chain-drive paddlewheel to move the vehicle. Herman Alkire of Adel, Iowa, used a motorcycle with runners front and back and a driving wheel in the center. Minnesota inventors Frank and Howard Sawyer and Iver Holm built "snowmo-cycles." E. A. Remezy of L'Isle-sûr-Sorgue, France, received a U.S. patent in 1913 for his auger-propelled automobile on runners. In the 1920s and 1930s the Tucker Motor Company developed an auger-powered snow vehicle.[10]

The Model T, with its simplicity, frame strength, and stature as a symbol of the glories of mass production, served as the basis for several attempts to manufacture a snow sled. In the 1920s Gilbert King and Frank Novak of Princeton, Wisconsin, built half a dozen Model T half-track snow vehicles. In 1928 Admiral Richard Byrd took to the Antarctic a Model T that the Ford Company had fitted with a track similar to that of a caterpillar tractor; he abandoned it 75 miles outside of camp. In 1939 he left behind a Snow Cruiser, a very large twin-tracked vehicle.[11] Businesses also sold various ski kits, paddlewheels, and track conversions for motorcycles and automobiles. In 1921 Charles Young of Maine filed a patent for a motor-driven sled. This vehicle "had an endless belt on an independent rear suspended power unit and a curved front with skis for steering" that anticipated the snow toboggan of Carl Eliason.[12] Also anticipating the Eliason machine, in 1917 Minnesotan Otto Johnson built a one-man motorized toboggan. Mike Bosak of Manitoba, a farmer and cabinetmaker, worked on a motorized toboggan in the late 1940s, in part inspired by the efforts of J. Armand Bombardier (see below) and others. Over the next several years he produced up to fifty machines, some of which members of the Antique Snowmobilers Club still have running.

Many aficionados consider Carl Eliason of Saynor, Wisconsin, to be the father of the motor toboggan. His vehicle may have been the first "mass-produced" snow sled. Eliason used a front-mounted water-cooled outboard motor. He saw the commercial possibilities of snowmobiles helping people through the hardships of winter, especially trappers, hunters, woodsmen, and sportsmen. He obtained a patent in 1927 and built forty to fifty machines in the garage behind his store, with no three exactly alike.[13] In 1941 he received an order for 150 of them from Finland. Realizing that he could not produce them all himself, he sold the manufacturing rights to the Four Wheel Drive Auto Company of Clintonville, Wisconsin. Eliason remained as chief consultant. The Clintonville FWD plant built four different models. Although the Finnish deal fell through, the U.S. Army later purchased 150 all-white Eliason/FWD "snowboggans" for use in the defense of Alaska.[14] Several FWD snow toboggan models followed.

During the 1920s Virgil White, a Ford automobile dealer in West Ossipee, New Hampshire, produced about twenty-five thousand "snowmobile" conversion kits adapted from Model T Fords. In 1913 he had invented a caterpillar-type tracked vehicle that could be mounted to the rear axle of a Ford with a set of skis for the front end. He referred to the vehicle on his patent application as a snowmobile. A snowmobile race on January 31, 1926, on Rangeline Lake near Three Lakes, Wisconsin, among drivers operating White's Model T snowmobiles may have been the first of the hundreds of competitions now held annually. White's snow vehicle was literally a Model T, with double rows of rear tires affixed with beltlike chains and skis up front for steering. White's vehicles were much larger and slower than modern snowmobiles, but the moving belt for traction and skis for steering remain the principle of operation. With a 20-hp engine and so much weight, the White snowmobiles moved slowly. Still, in the absence of plowed roads they did the trick and were quite popular.

Several inventors chose the air sled design for their snow vehicles. Gene Schnaser's examination of Patent Office materials indicates that at least half a dozen air-powered snow vehicles were developed in the first half of the twentieth century. One was the Aero-Sled, publicized in the January 1917 *Motor Mechanics.* The editors described the Aero-Sled as "'a handsome affair' with curved dash, soft seats, regular auto steering, making it virtually an automobile on runners, capable of speeds up to 45 mph and even faster." Felix Hakkinen, a supply sergeant of Company F, Seventh Infantry, in Haines, Alaska, built an air sled in 1937. At first he used a front prop, but the snow that it blew into the face of the operator made the trip rigorous to say the least, so Hakkinen turned to a rear mount. He employed a 25-hp motorcycle engine that ran at 2,500 rpm and produced blue flames.[15]

Another player, Glen Gutzman, built an air-powered sled that sold well in the 1960s. Informally called Trail-A-Sled (later Scorpion, Incorporated), Gutzman's company set up shop in Eagle Bend, Minnesota, to produce a series of aluminum air sleds. Each machine improved somewhat on its predecessor, and by word of mouth this triggered sales that led to a relocation of manufacture to a larger garage in Crosby, Minnesota. There Gutzman joined forces with Crosby residents Richard and Eugene Harrison in pursuit of a fiberglass model. The three men formed Trail-A-Sled. The firm moved ahead with fiberglass designs produced out of a small garage. A new air sled with a Lycoming 125-hp motor appeared in 1960, with fifty turned out by 1963. The sled was sold generally for commercial purposes and required good conditions to achieve maximum performance. "Operation in tight quarters was out of the question and deep snow and drifts posed serious problems," but in optimal conditions the "air-sled ran like a dream and approached speeds of 100 mph."[16]

Inspired in part by Bombardier's Ski-Doo, Trail-A-Sled also pursued crude versions of tracked snow machines. The first prototype appeared in 1961. The snowmobile used fiberglass and plywood construction, a rudimentary cleated track, and fiberglass-treated two-by-fours in the suspension. While selling only one dozen of the snowmobiles, Trail-A-Sled got orders from Polaris Industries for parts (backrest, console, fenders) for a new snow machine called the Comet. As orders grew, Trail-A-Sled decided to drop its air sled and gear up for a tracked snowmobile of its own, called the Scorpion. The company produced nearly fifty machines in the fall of 1963 for the 1964 model year. The machine's cleated-track system was noisy and cumbersome. This led to the development of a track system using vulcanized rubber and mesh fabric over steel chain, the first of its kind patented in the United States.[17]

Like Bombardier and other snowmobile pioneers, Gutzman was initially forced to be part designer, part builder, part salesman. He put a Scorpion in his Volkswagen and drove around the United States and into Canada visiting "small town mechanics, chainsaw dealerships—anyone who might be interested in expanding their winter business." This led to orders for scores of Scorpions even through such retailers as Sears and Roebuck. The Scorpions gained national attention in trade magazines including *Mechanix Illustrated.* In 1964–65 Trail-A-Sled produced five hundred Scorpions that came with various engine sizes, could reach a speed of 40 mph, and could navigate a 50 percent grade. The firm manufactured its own hoods, windshields, tunnels, seats, engine mounts, clutches, and bogie wheels and hired subcontractors to handle metal machining (including clutches and axles), vulcanized products, welding, fabrication, and chrome plating. The firm produced 2,150 snowmobiles in 1966, 5,000 in 1967, and 8,000 in 1968, when Gutzman was named Minnesota's Small Business Man of the Year.[18]

Misfortune—in the form of a fire and uncertain finances—led to Scorpion's sudden decline. The owners moved operations to a city-owned arena where 17,500 square feet of unused ice-skating space became an assembly plant. A devastating fire destroyed that production facility, but this led the company to move to a new 70,000-square-foot complex that could manufacture up to fifty thousand machines in a season. In 1969 the firm prepared to produce twenty thousand units, getting close to the goal of two hundred units per day and now employing nearly three hundred persons. The firm announced plans to produce thirty thousand units for model year 1970 and fifty thousand for model year 1971. But the privately held company could not easily finance the production growth needed to expand market share. It also faced a number of larger and better-established firms that could offer buyers such attractive inducements as installment purchasing. When Fuqua Industries of Atlanta, Georgia, made an offer to purchase Trail-A-Sled, the owners sold out, staying on as managers.[19] Scorpion production was ultimately bought up by Arctic Cat, which closed the plant and moved production to its Thief River Falls facility.[20]

Bombardier in Rural Quebec

J. Armand Bombardier of Valcourt, Quebec, who created the world's largest snowmobile company before his early death in 1964, advanced his snow machines at roughly the same time as Eliason. He worked first on air-propeller-driven machines made from a car frame, engine, and accessories similar to the Model T half-tracks that preceded him. In the winter of 1922 he built a motorized sleigh, using a rear-mounted motor from an old Model T and a frame that had four ski runners. The vehicle was steered by a rope. In 1927–28 he turned to tracked machines. But, as Schnaser points out, Bombardier "started to realize that the steel belt was too heavy, could not last long, and harmed the tires," and he began testing conveyor-belt tracks.[21]

In 1934, having established himself as a successful repairman of automobiles and agricultural equipment, Bombardier embarked on the quest for a true snowmobile. His son had died of appendicitis that winter, in part because emergency medical help had been unable to reach him in time over the impassable snow-covered roads. To meet such needs, Bombardier set about designing a motorized vehicle light enough to travel on snow, with a motor, traction, and suspension adapted to the changing consistency of snow. He had some success in developing prototypes but also incurred criticism from family, friends, and outsiders for the costly failures he built along the way. Automobile engines clearly were too heavy for a snowmobile, so he built a lighter 45-kg motor. That engine tended to overheat, however, leading Bombardier back to automobile engines and a heavier snow vehicle.

In the year after his son died, Bombardier perfected a design using toothed wheels covered in rubber, with a rubber and cotton track that wrapped around the back wheels. According to the Bombardier Museum website, "This revolutionary sprocket and track system [was] at long last the solution for snow travel."

Bombardier established a small plant in Valcourt in which he produced seven snowmobiles during the winter of 1936–37, a model called the B7 (meaning that it had room for seven passengers; it was more a snow coach than a snowmobile). He received a patent for the design in June 1937. Bombardier observed that snow and ice accumulated in the vehicle's wheel spokes. He solved the problem by assembling a press to manufacture solid wheels, which appeared on the 1940 models.[22]

Bombardier produced various tracked vehicles during the 1930s and 1940s, but as historian Lenny Reich writes, the "bottom dropped out" of the market for such vehicles when "Canadian provinces began to plow their rural roads." Only in the late 1950s, when a northern Canadian missionary asked for his help in traversing the winter snows of remote areas, did the snowmobile as we know it begin to develop, in the form of the machine Bombardier called the Ski-Doo. There were other snowmobiles available at the time, but they were much more cumbersome to maneuver than Bombardier's Ski-Doo. The Ski-Doo's strength was its maneuverability, with the tread placed back and the skis set forward and to the sides wide apart. The other machines used pairs of long fixed skis and placed the drive tracks between them. They didn't bog down in snow, but they were hard to steer. Bombardier knew how to optimize track size relative to vehicle weight and to the ratio of track width to length.[23]

Bombardier drove around Quebec Province trying to sell his sleds. He often parked near local newspaper offices to ensure free publicity. This marketing strategy generated so much demand that he had to build a new plant with an annual production capacity of two hundred vehicles; it came on line in January 1941.[24] In 1941–42 Bombardier developed the B12, a twelve-passenger vehicle with a longer, more aerodynamic profile than the B7. Orders increased until Canada declared war on the Axis powers. Material and manpower rationing now prevented manufacture of civilian vehicles, but Bombardier's offer to the Minister of Munitions and Supply to join the war effort led to a contract for a prototype military snowmobile troop transport for use in such snowbound theaters as

Norway. Starting from the B12, Bombardier designed the prototype B1. The Canadian forces ordered 130 vehicles, to be delivered in four months. Since his Valcourt facility was too small, Bombardier shifted production to an existing Montreal factory, although he continued to manufacture parts in Valcourt to maintain local employment. Bombardier turned out over 1,900 armored tracked snow vehicles for the military between 1942 and 1946, while his civilian production grew from 27 units in 1942–43 to 230 units in 1945–46.[25]

The postwar years were boom years for Bombardier. Demand for civilian snowmobiles increased rapidly, as did the company's revenues. Through his years as founder, director, chief visionary of the company, Bombardier showed ingenuity, flexibility, and aggressive innovation in the development of new markets. In 1947 he built an assembly-line plant with a capacity of one thousand vehicles inspired by Ford factory assembly lines. The B12 enjoyed great popularity in a range of markets including public and materials transport, ambulance and rescue services, communications and electricity companies, mining and prospecting, and transport of missionaries to Inuits in the isolated Canadian north. Between 1945 and 1951, Bombardier sold nearly twenty-six hundred machines. The C18, seating eighteen adults or up to twenty-five children, found popularity in Quebec and Ontario as a "school snowmobile." In 1947–48 Bombardier's sales reached $2.3 million, ten times the figure for 1942–43.[26]

By the late 1940s, however, North American governments had recognized the need to keep highways and roads clear in the winter. In 1949 the Quebec government adopted a policy requiring rural routes to be cleared. This was a major blow to the snowmobile market. In one year, sales fell nearly $1 million. Bombardier took this as a challenge to expand his product lines to include ATVs, industrial trucks and tractors for the forest industry, all-track vehicles, sidewalk sweepers and plows, and vehicles equipped with an interchangeable system of wheels and skis. The TTA (Tractor Tracking Attachment) improved traction in muddy and swampy terrain, enabling Bombardier to find a market with tractor manufacturers in North America, Europe, and South America.[27] Bombardier considered the Muskeg tractor his greatest success. The Muskeg was an all-track, low-impact all-terrain vehicle. First produced in 1953, it met demand for construction and transport in all sorts of terrain, from snow to swamp to desert, and found a market around the world.[28] Bombardier skidders and delimbers also made their way deep into the forests of industrial lumber operations.

Given his success at developing new markets and building flexible production facilities, it is no surprise that Bombardier set the standard for the personal snowmobile industry with the Ski-Doo. Mass production of the Ski-Doo began in the autumn of 1959. Missionaries, trappers, prospectors, surveyors, and other people isolated by winter snows embraced the machine, as did sports and outdoor recreation enthusiasts. Production increased rapidly, from 225 snowmobiles in 1959–60 to 8,200 in 1963–64, with the rapid growth requiring several expansions of the Valcourt facilities. In 1967 Bombardier sold forty-four thousand units, forty other snowmobile makers had entered the market, and the total number of snowmobiles in North America had reached two hundred thousand.[29] After J. Armand Bombardier's death in 1964, first his son and later his son-in-law took over the company, taking it into the twenty-first century as a major player in the Canadian transportation, forestry, and military industries.[30] Today the huge corporation produces rail vehicles including trams, light rail and metro cars, regional, business, and amphibious aircraft, and a number of other products.

The first Ski-Doos were simple, boxy, utilitarian vehicles. The operator's manual included easy-to-read diagrams so that the owner could repair the drive or transmission himself, a relatively common occurrence. The 1961 Ski-Doo weighed 385 pounds and was powered by a 7-hp two-stroke JLO engine, capable of moving the machine "through woods, hill and dale, crossing obstacles such as snow covered ditches, mounts, hollows, snowbanks, etc." Improvements in suspension in 1961 included leaf springs replacing the original coil springs, which helped steering. Advertisements showing families operating the vehicles (without helmets or special protective equipment) promoted the Ski-Doo as "enjoyable for all ages."[31]

Bombardier made the 1963 Ski-Doo BR series more sleek, with a reinforced fiberglass cowl, increased traction, larger track-bearing area, increased cruising range, and a top speed of 35 mph, while the RD series had two tracks and, at 377 pounds (nearly 150 pounds heavier than the BR), could muscle only 20 mph. The improved 9-hp engine vibrated less and had a better starter. Promotional brochures clearly targeted outdoor enthusiasts as well as families, with photos depicting rope-pulled snow-skiing, for example, or a father and son riding together (the man without a helmet, the boy riding in the front carrying rack).[32] By 1969 the Ski-Doo had become today's snowmobile, with complete instrumentation in a wood-grain dashboard, an electric starter, even an automatic cigarette lighter. Two-stroke single- and twin-cylinder engines were available for most models. There were three series: the Nordic, Olympic, and Alpine. The Nordic 371 had an 18-inch track, for more bite and grip in deep snow, and a 371-cc Rotax engine to get you through it. The T'NT 399 and T'NT 669 were ski racers, the former with a 399-cc twin-cylinder two-stroke engine developing 30 hp, and the latter a 669-cc twin-cylinder two-stroke engine developing 45 hp. A new muffler and frost-free cables were welcome improvements. The Ski-Doo Plaisted Polar Expedition in 1968, in which Bombardier snowmobiles "with light-footed action" crossed the Arctic Ocean "on cracking, moving ice floes," carrying men and equipment 825 miles, enabled the company to tout the machine's rugged dependability.[33]

Another short-lived competitor in the early market, the Hus-Ski, used a twin-track tractor with a ski sled pulled behind for the operator and passengers. The operator steered the vehicle using handlebars on which were mounted a brake and throttle lever that extended from the tractor. The passenger placed his feet in tow clips on the skis. The Point Claire, Québec, facility turned out the first Hus-Skis in 1962. Hus-Ski changed models and features season to season, adding larger engines. The crucial innovation was dropping twin tracks in favor of a single wide-apron track. Also, a side drive sprocket assembly replaced a down-the-center drive. A system of staggered bogie wheels that supported the wide track replaced the wooden slide rails. But this led to frequent tip-overs, so the units were later fitted again with twin tracks. In 1965 Food Machinery Corporation bought Hus-Ski to add a winter product, a snowmobile, to the summer line of lawn care products produced by its subsidiary, the Bolens Company, of Port Washington, Wisconsin. Bigger engines, an electric starter, saddle bags, clamp-on windshields, and other changes carried the newly named Diablo Rouge until 1969, when Bolens turned to a more conventional sled called the Bolens Spring 620.[34]

According to a 1960s advertisement for the Hus-Ski, the machine's advantages were maneuverability, dependability, portability, and ease of maintenance. A "spacious deluxe seat" enabling riders to sit upright in a "normal and uncramped position" guaranteed comfort. The steel frame ensured durability, but "necessary repairs may be made easily and inexpensively." There was "no need to worry about messy oil changes or worry about engine . . . lubrication. Oil is mixed with the fuel by the dealer or filling station attendant." The advertisement claimed that the Hus-Ski was safe even for children, though as we shall see, such claims clearly did not take into account the inherent risks involved in riding high-speed machines without roll bars or passenger restraints.

The Northern Minnesota Contribution: Polaris and Arctic Cat

Two other heroes in the history of American snowmobiles were Edgar Hetteen and David Johnson, proprietors of a small fabrication facility in Roseau, Minnesota. The Hetteen Hoist and Derrick company repaired farm equipment, helped the Rural Electrification Administration with pole setting, and produced one-of-a-kind machines. Hetteen built the forerunner of the Arctic Cat, the Model 100 "Tin Lizzie," in 1952. According to C. J. Ramstad, the business was a "big struggle," a "hand to mouth" affair, until 1954, when the company built its first mulcher. But the mulcher was a seasonal product, and "people had an interest in hunting in the winter." In 1955 a neighboring farmer asked Hetteen to build him a gas-powered sled. The company was already building such specialized farm implements as straw cutters and post setters, but this was a new challenge. Hetteen set out to design some sort of tracked vehicle steered by skis. He built two versions, the second better, but material and fabrication costs made it too expensive to produce a series of them for market sales. Over the next few years, however, he learned how to build less expensive and more reliable machines, and eventually he abandoned fabricating farm equipment. The company produced five sleds in 1955–56, seventy-five in 1956–57, and three hundred in 1957–58.[35]

Meanwhile, when Edgar Hetteen went off for two weeks in 1956, his partner David Johnson built a snowmobile. The Iron Dogs, as he called them, caught on quickly. Johnson's son recalls that "Dad once got a happy letter about the wonderful machine" from a man who called it "great for me and for my family." The man had gotten caught in frigid cold. "I lost my foot," he said, "but love the machine."[36]

Johnson encouraged Edgar and his brother, Allan, to work on a snow vehicle. They took some inspiration from the Eliason motorized toboggan. Their first machine had an auger as the propulsion device, and they nicknamed it Screaming Lena because it snaked sideways when it hit hard-packed snow or road surface. This experience led them to a tracked vehicle that used an elevator chain. They sold the first one to H. F. Peterson, a lumberman and avid hunter, who took the vehicle into the woods in search of rabbit and fox. When the snowmobile broke down, as frequently occurred, Johnson and the Hetteens had to drag it manually out of the woods to the nearest road to bring it back to the shop. From their repairs they got ideas for improvements as well as for other commercial possibilities. Throughout the early years they maintained the vision that the Sno-Traveler was first and foremost a utility tool, not a recreational machine. It would assist utility workers, lumbermen, and trappers and hunters. Johnson and the Hetteens also envisioned Inuit people and missionaries using the

Sno-Traveler instead of dog sleds and doctors using it as an emergency vehicle. To promote the vehicle, they took Sno-Travelers to The Pas, Manitoba, for the annual trappers' festival. When the vehicle won a race with dog teams, it captured local imaginations. The Winnipeg distributor immediately placed an order for twenty-five snowmobiles.

When pressed by his board of directors to abandon snowmobiles in favor of the other fabrication activities that were the bread and butter of the facility, Hetteen, who was determined to market the snowmobile, settled on a public-relations event to highlight the machine's promise as a workhorse. On March 5, 1960, he and his wife, Bessie, embarked on a 1,200-mile trip across Alaska, accompanied by Earling Falk (an employee) and Rudy Billberg (an Alaskan bush pilot). Unlike the machines of today, the early Polaris snowmobiles they took with them were wobbly and underpowered. They started poorly, often broke down, and needed frequent repairs. The team had two 10-hp Sno-Travelers and one 7-hp Trailblazer. With gas cans and snowshoes strapped on, the snowmobiles labored to tow two freight toboggans, each carrying 900 pounds of supplies. Bessie either doubled with one of the men or stood on one of the toboggans, which made the trek even harder. High winds and temperatures to −40° further hampered the travelers. When temperatures rose, they faced other problems: thin ice that prevented them from moving along the rivers and lakes they preferred because of the smooth surfaces. During the 1960 Alaska tour, Johnson steered his snowmobile with ropes from behind the machine, against the chance of the snowmobile going into a crevasse. Repairs were a nightmare: "During one major repair session, Edgard repaired his machine with such cold, numb fingers, that he dropped his hammer into it. Without the protective shroud in place, the running motor sheared all of the cooling fins from the flywheel." But the snowmobile ran well anyway. The team took twenty-one days to cross 1,200 miles of the Arctic, from Bethel on the Bering Sea coast to Fairbanks. At times they traveled at a rate of under 8 mph.[37]

The trek was a public-relations success. Airplane and ham

radio covered the wilderness journey. One of the drivers referred to the machines as "cats" in his diary, on the radio, and then in *Field and Stream,* and the name stuck. They became "Arctic Cats." The publicity success enabled Hetteen to secure financial backing from a supermarket owner in Thief River Falls. He set up shop in a food warehouse (where he also built portable steam-cleaners and black-light bug killers). In December 1961 Hetteen's first snowmobile, the Polar, appeared. He produced twenty of them, then changed the name to Arctic Cat. The snowmobile had handlebar steering and ski/wheels, could traverse swamp, marsh, and bog, had a 9.5-hp engine, and cost $1,210. The company targeted conservation, forestry, telephone, power, and light companies for sales.[38]

The period 1955–65 determined what the snowmobile would be. Johnson's son, Mitchell, described the two-trunked family tree that comprised the snowmobile industry: Bombardier and Arctic Cat / Polaris. After a falling out, Hetteen left Polaris to found Polar Manufacturing in Thief River Falls, Minnesota, which in 1962 became Arctic Enterprises. Arctic produced the Arctic Cat. Competition between the two families and other manufacturers led to a series of important innovations. Would the snowmobile be rear or front engined? If rear, the passenger would be warmer, but if front, the vehicle would be easier to turn. How could economies be achieved in manufacture? A unitized body was the key. How might the vehicle's weight be distributed? Concentration of the center of mass would make the vehicle more fun to drive.[39]

During the 1960s Arctic Cat successfully introduced a series of new models, all with two-stroke engines, while expanding the product line. The Arctic Cat 100 (1962) employed a front engine and used a bicycle headlight that often fell off; it ran through a generator turned by a rubber wheel contacted to the primary clutch. The company introduced bigger, heavier sport and workhorse models. As part of its growing publicity campaigns, the company published a newsletter, the *Arctic Cat Howl,* and in 1963 began to sponsor racing derbies. In the 1963–64 season the company offered thirteen models (some with electric starters, some with front engines, some with rear engines, some with headlights but rudimentary if any suspension, and so on), up from six. Fiberglass had begun to replace metal in the sleds. Production reached 803 machines in 1964, and in 1965 shipping to foreign countries commenced.[40]

In 1967–70 Arctic Cat offered the Black Panther with the improved suspension, a flip-top hood, and a riveted aluminum tunnel. The company targeted families and "outdoor wilderness seekers" and introduced lines of "Arctic Wear" including jackets, hats, gloves, and boots. Kawasaki single-cylinder 292-cc engines powered the Cats. Sales reached twelve thousand in 1968, and then a sales boom hit both the company and the industry as a whole.[41] In 1970, when production reached fifty-five thousand units, the company acquired a new 440,000-square-foot manufacturing plant to meet demand. During the 1970s Arctic Cat secured a significant share of the market. Snow derby competitions with Ski-Doo, Polaris, Rup, Sno*Jet, and Moto-Ski helped push industry sales to 420,000, with Arctic Cat turning out 100,000 in 1972 and 99,000 in 1973.[42] Sales of clothing, trailers, sleds, gloves, goggles, and boots grew substantially as well. The snowmobile ended the "hibernation of towns, hotels and resort areas in winter."[43]

Polaris too expanded rapidly in the 1960s, in 1964 moving into new 47,000-square-foot plant with increasingly modern assembly lines. In 1964 the machines also went from cleated (metal) to all-rubber tracks; the new, more powerful engines would have ripped the cleated ones apart. In 1968 the plant reached 120,000 square feet, with a "Detroit style assembly line" and nine hundred employees producing five models of snowmobile: the Playmate, Charger, Mustang, Voyager, and TX.[44]

From "Snofaris" to Industry Consolidation

The three crucial components of the snowmobiles were their engines, clutches, and suspensions. The engines were unique to each company. Since Honda is an engine company, its engines were (and remain) its snowmobiles' trademark. For Polaris, the suspension and clutch were distinctive. The Polaris company's constant variable transmissions have been adopted by all manufacturers. Polaris, Mitchell Johnson told me, *is* suspension. The suspension—its travel, response, and so on—was designed to reduce weight and get the machine through any kind of snow. Polaris "differentiated itself in customers' minds by starting faster, running quieter and lasting longer."[45]

Although snowmobiles had first attained commercial success among police, repairmen, trappers, doctors, farmers, and mailmen, recreationists now became their greatest fans. Two thousand clubs scattered throughout the country served members with advice. The clubs sponsored scores of meets, "snofaris," "snodeos," and the like. In an effort to build ridership, the companies also promoted races and employed their own daredevil test-riders. The clubs and manufacturers held roughly a thousand races and rallies annually, each calling itself the "largest," "richest," or "most unique." Alaska's Midnight Sun 600, the "coldest and cruelest" of them all, tested machines and operators more severely than any other. In 1969 more than three hundred starters representing twenty manufacturers set out. They raced at speeds up to 90 mph, at elevations up to 3,000 feet, and amid snow gusts of 70 mph that blew some of them off roads and knocked others out of commission. Subzero temperatures made repairs impossible. Frostbite claimed dozens of racers. One driver rigged his wife's electric hair drier to his helmet for warmth. At the end of day one, only eighty-one riders were left. Temperatures on day two dropped to −70°, and freezing exhaust reduced visibility to 50 feet. Only thirteen drivers and vehicles managed to finish.[46] Other races included the 468-mile Winnipeg to St. Paul trek, held one year in −25° temperatures.[47] The races of the 1970s were certainly a marketing ploy, but they also helped manufacturers to improve their engines and suspension. With the development of front independent suspensions, manufacturers moved full scale from racing machines into consumer machines.[48] In 1976, snowmobiles appeared in a segment of the *Donny and Marie* television show, perhaps the first time they had been used as a prop on American TV.

The 1969 *Field and Stream* snowmobile buyer's guide listed fifty manufacturers and four hundred models. In 1968, these manufacturers had sold 285,000 units. They anticipated a 23 percent increase in 1969. A number of them were involved in garden tools or recreational boating as well (Homelite, Evinrude, and Yard Man). Styling had begun to receive more attention, with such extras as toolkits, spare parts, and instruments becoming common as standard equipment or factory options. Slide suspension systems replaced bogie wheels in some models. Bigger engines and higher horsepower were the rule. The *Field and Stream* editors noted that not all manufacturers were entirely forthcoming in their brochures, descriptions, and specifications: "Hedging, corner cutting and out-and-out falsification of figures (hp, ground pressure, heights) by some manufacturers has been so flagrant in past years that some companies now simply refuse to disclose this information."[49]

At the peak of the industry in the early 1970s, nearly two hundred companies manufactured snowmobiles. Some of them (e.g., Polaris, Arctic, Lionel, Scorpion, and Moorehead Plastics) built snowmobiles for other companies to market under their own names. Several manufactured tracked vehicles that were more an all-snow vehicle than a snowmobile.[50] Other companies sprang up to provide spare parts. According to a 1970 price list, the Gates Corporation—still in the snowmobile v-belt market—alone produced over fifty different belts for eighty-six snow vehicle manufacturers.[51]

The industry peaked during the three-year period 1971–73, when one hundred manufacturers turned out 1.8 million units. Then an industry shakeout occurred that left only thirteen companies still building sleds in 1975. The major factor in the shakeout was the saturation of the market by firms without the R&D foundation or product support to keep customers happy. Another was the OPEC oil embargo, which drove gasoline prices up, combined with bad weather (not enough snow). By 1980 there were only seven manufacturers, and as of January 1, 1980, only 146,000 units had been sold during the 1979–80 winter season, while another 150,000 sat in stores, at an average price of $3,000 each. Rebates and premiums had become the industry norm. By the late 1990s the industry had rebounded, producing 225,000 units in 1996–97.[52]

Today there are only four major manufacturers of snowmobiles in North America: Arctic Cat of Thief River Falls, Minnesota; Bombardier Recreational Products, of Valcourt, Quebec; Polaris, of Medine, Minnesota; and Yamaha, of Cypress, California. In 2005, manufacturers sold over 100,000 new snowmobiles in the United States and 46,304 in Canada at an average price of nearly $8,000. Almost two thousand dealers in the United States, nine hundred in Canada, and four hundred in Scandinavia sold the vehicles. Owners have registered approximately 2.6 million snowmobiles worldwide, with 1.6 million in the United States (see table A.1), 760,000 in Canada, and 318,000 in Scandinavia. Minnesota, Michigan, and Wisconsin combined account for 56 percent of U.S. snowmobile registrations. World annual snowmobile sales over the last fifteen years have ranged between 150,000 and 260,000 annually (see table A.2). The International Snowmobile Manufacturers Association calculates the economic impact of the snowmobile industry to be $21 billion annually in the United States, $6 billion in Canada, and $1.6 billion in Scandinavia, with the industry generating eighty-five thousand full-time jobs in North America alone.[53]

Who owns snowmobiles in contemporary America? The average snowmobile owner is a married man forty-one years old with an annual income of $70,000. He rides 990 miles per year and spends $4,000 annually on snow recreation, tourism, and related products. Two-thirds of snowmobilers trailer their vehicles. Snowmobilers have 230,000 miles of groomed and marked trails in North America and two thousand clubs striving to open up even more miles. This is a market that manufacturers do not want to see restricted by regulation stipulating where operators can ride or imposing new safety and noise standards that might make machines more expensive or change the ride experience. The manufacturers' association points out that they raised over $3 million for charity in 2004–5; whether this indicates a caring attitude is unclear, since this amounts to only about $2 per registered owner per year.[54]

The introduction of snowmobiles and other recreational machines triggered a demand for personnel trained in the development, design, maintenance, and servicing of these vehicles.

In 1968, Gogebic Community College in Michigan, one of the major centers of snowmobile activity, began offering vo-tech training in small-engine and structural concepts technology. Students earned associates' degrees in applied sciences to meet the demand for dealer technicians and other positions connected with snowmobiles, outboard-inboard motors, chain saws, snowblowers, and other engines. According to Gogebic's promotional materials for 1968, "The snowmobile is an excellent example of small engine application. Its growth in the snow-belt states as a recreational and sports vehicle is almost unbelievable. North American sales of snowmobiles was a mere 15,000 vehicles four years ago. This year sales are expected to top 200,000 units. Sales for 1969 are forecast at 500,000."[55] Gogebic's program offered students access to approximately $18,000 worth of vehicles, small engines, equipment, and instructional aids made available to the program on loan or by outright gift from manufacturers. The program of instruction included training on engine fundamentals and repair; carburetion-ignition systems; drive systems and brakes; chassis design; tuneups; field and lab testing; fuels and lubricants; clothing and accessories; and so on.[56] To this day, Gogebic Community College offers courses on various aspects of management and service for the ski industry.

The close relationship between higher education and snowmobiling has evolved through formal competitions that encourage leading associates and engineers to make improvements in the technology. In 2005, Arctic Cat, BRP (formerly Bombardier), Polaris, and Yamaha sponsored the Society of Automotive Engineers (SAE) sixth annual Clean Snowmobile Challenge at Houghton, Michigan. Over 130 students from various U.S. and Canadian universities participated in the competition, held at the Keweenaw Research Center, which is affiliated with Michigan Technological Institute. In the competition, novice engineers presented thirteen projects. The University of Wisconsin, for example, designed and built "a clean, quiet, high performance snowmobile" for use in environmentally sensitive areas.[57] The university-built snowmobiles competed in a 100-mile endurance run, a fuel economy test, and a new rider comfort evaluation to measure shock and vibration over a bumpy course. First place went to SUNY-Buffalo, Clarkson took second place, and Wisconsin-Madison took third. SAE has designated Michigan Tech to continue coordinating the annual Clean Snowmobile Challenge.[58]

Snowmobiles and Cultural Change

In 1969 the editors of *Holiday*, a magazine devoted to encouraging travelers to visit new places, lauded the snowmobile's contribution to winter holidays. Snowmobiles "creep, crawl and can nearly fly across the snow at speeds up to . . . 95 miles per hour. They go cross-country . . . for four hundred miles. They tow skiers, carrying whole families in tandem trailers on skis. They congregate by the thousands on winter weekends for touring . . . and show up for snowmobile races by the tens of thousands." The editors continued, "The snowmobile does for the woods what the outboard did for the lakes. A snowmobile can take an otherwise sedentary fellow who might normally be tempted to spend a thousand dollars flying himself . . . to Florida for a week, put him on the snow for the same price and make him love it." Snowmobiling was relatively comfortable, speedy, easy to do (like driving a car), and open to all weather and terrain. And it was family fun.[59]

Yet the family fun also involved noise, pollution, and accidents that increasingly caught public attention. The nation's seven hundred thousand snowmobiles and the $400 million snowmobile industry became a hot subject in state legislatures as "irate homeowners, nature lovers and safety experts denounce the vehicles' hedge-chopping, noisemaking and accident-prone proclivities." As both club members and opponents noted, snowmobiles could be operated recklessly and quickly, or sedately and slowly. The snowmobile helped fight the isolation of the winter and stimulated business at small local northern stores, hotels, and restaurants. Towns attracted the snowmobilers by building bridges, cutting and grooming trails, printing maps, and holding festivals. Technological improvements permitted smoother ride, better control, and higher efficiency; new alloys reduced the frame's flexibility and lowered the center of gravity while improving strength and weight distribution. Yet these technical improvements also made the high-speed whine and blue smoke of two-stroke engines ubiquitous in snowbound forests and fields.[60]

Reviewers of snowmobiles always praised them for their simplicity, for the sheer fun they offered, and for the way they facilitated access to the woods. The government also contributed to their spread through the woods. "National forests and parks are marking trails for snowmobiles," one magazine reported. There were few laws prohibiting their use. The magazine urged operators to bring along emergency equipment and adequate clothing, stay on trails, leave a route plan, and never travel alone.[61] Unfortunately, not everyone takes this advice, and few people have considered the way in which snowmobiles, like other technologies, have unanticipated impacts not only on wildlife but on human communities.

Technologies reflect the social, political, and cultural institutions of their creators and users. Technologies are also a force of social, political, and cultural change. Perhaps the most vivid example of this is the modern factory, whose rise during the Industrial Revolution contributed to urbanization and the formation of the working class. In a series of overstated yet compelling investigations, the historian Lynn White Jr. makes the case that the stirrup triggered the rise of feudalism. He argues that eyeglasses altered the competition between established scholars and novices for the few spaces available in the universities of the Middle Ages by allowing elderly scholars to continue to read and write. He contends that the chimney contributed to class differentiation by allowing peasants to leave the manor to live in their own small hovels with a flue. American observers have long argued that the telephone, the automobile, and the computer have changed the way we live and work, court and marry, and communicate.[62]

The snowmobile has had a significant impact on the way people in continental climates spend their winters, earn their income, hunt, relax, and celebrate. Tourist economies of winter resort areas now rely heavily on the machines. The snowmobile has had such positive effects as enabling people to overcome the isolation imposed by deep snows and heavy freezes and facilitating rescue by medical personnel and law enforcement officials. Many hunters have abandoned snowshoes in favor of machines. Snowmobiles have accelerated change in the fabric of life in some communities, just as they have interrupted snowshoeing, winter hunting, and cross-country skiing with noise and pollution. Snowmobiles, along with other such machines with small-bore engines as boats and ATVs, have altered the way Inuit, Sami, and other northern people live, in some cases pulling their economies apart, creating new dependencies on parts and oil, and interrupting traditional ways of hunting, not to mention generating new concerns about public health and the environment.

The Sami and Inuit have developed a stable lifestyle based on hunting and herding. Entire communities work together to share the difficulties, costs, and dangers of their precarious economy. They also share the benefits, harvests, and celebrations of life. When snowmobiles and other machines appeared in those communities, suddenly one or several men might command enough resources to hire others, out-compete others, and challenge local lifestyles even to the extreme of overhunting. Not surprisingly, many villages sought to save money to buy snowmobiles themselves. Some turned to debt financing. All relied more heavily on such petrochemical products as oil and gas and automotive products such as spare parts. Their buying and selling patterns came to involve new markets whose successes meant little for local people and whose profits were rarely plowed back into the community. Snowmobiles enabled Inuit, Sami, and others to gain access to new, unfamiliar products including alcohol. They triggered changes in work habits and skills. Many Inuit parents lament that children today know much less about local climate, flora, and fauna because the machines enable rapid access to the wilderness.

The Sami, a population of roughly forty thousand individuals who live in Norway, Sweden, Finland, and Russia, make their living from reindeer herding, some fishing, and some agriculture. Their traditional diet was the meat, blood, and inner organs of reindeer, cereal products, and to some extent fresh and dried fish and berries.

This diet reflected the Sami lifestyle and the local availability of food, but it changed significantly in the second half of the twentieth century as the traditional lifestyle changed. A major reason for the change was the introduction of snowmobiles, motorcycles, and other technologies that facilitated daily work and enabled management of herds with fewer moves. The dietary pattern of the Sami in northern Norway, where roughly 50 percent of all Sami live, became more typically Norwegian.[63]

The snowmobile contributed to the steady erosion of Inuit and Sami life in other ways as well. While many people celebrate the incorporation of traditional peoples into a more modern, consumerist culture, others contend that they—and we—are left the worse off for it. Soviet officials deliberately, and often violently, destroyed the lifestyles and cultures of the Nenets, Khanti, and others, outlawed their shamans, and collectivized their herds, to turn them into good Soviet citizens. Western governments have introduced school and economic development programs that have had the same outcome of cultural change, even if that outcome was unintended. Increasing reliance on money earned through jobs disrupted Inuit lifestyle, leading many of them to abandon collective hunting and the tightly knit bonds of community "based on kinship [to] ensure that food is distributed fairly," in the words of geographer George Wenzel. Increased contact with southern society and money led them to rely more on outsiders for essentials. Hunting became more expensive as "dogs and sleds became outmoded, replaced by powerful snowmobiles." The strains of buying, maintaining, fueling and repairing have been great. People disconnect their phones and go without other "necessities" to keep the snowmobile going.[64]

In response to the dislocations that followed the appearance of the snowmobile, the Canadian government intervened to create new jobs. Yet those jobs went to those few individuals who could read and write English, leaving unemployment at roughly 60 percent. In addition, those holding jobs were less likely to share their wealth with members of their kin than hunters were. The people found the jobs empty, requiring little skill, while hunting gave status that the jobs did not. Wenzel believes that with the creation of the government of Nunavut, based more and more on Anglo-Canadian notions of democracy, and with the creation of bureaucracies, more jobs will flow away from small communities and toward the bigger northern towns, and little money or support will flow back.[65]

The prime minister of the Nunavut government, Paul Okalik, worries that having a government and programs similar to those of other Canadian provinces has done nothing to stem the loss of traditional hunting knowledge. Okalik said:

> What is different about Nunavut is that we have a public government based on Inuit Qaujimajatuqangit, Inuit Traditional Knowledge. . . . We are struggling every day to find more ways to blend our past with our future. Our Elders are passing on their wisdom and knowledge to our Youth. We Inuit have accepted many modern trappings; the rifle and the snowmobile; the shack that became a small house that became row houses, apartments, and condos; the cotton, the duffel and now the gortex and fleece; but through it all we still maintain our Inuit spirit. Our art, crafts, carvings, hangings and prints imitate Inuit life of the past but bring it into the present.

Okalik adds that several technologies have passed from Inuit culture into European and North American culture:

> Our technological ability, driven by survival gave us the qayaq (kayak), the qamutiik, (a sled that is roped together for strength), the annuraaq, (anorak as you might know it, a pull over jacket for cold weather); your culture has taken these items and adapted them for your use. We are proud to share these items and only ask that you recognize where they came from.[66]

If only western technology had been so easily assimilated into Inuit culture. In the 2001–2 winter season the Northern Transportation Company supplied Nunavut with impure gasoline that damaged two- and four-stroke engines on boats and snowmobiles, requiring extensive repairs. The Nunavut government announced interim payments to owners for 75 percent of the repair costs. Owners had to submit a completed government-approved claim form to the Risk Management Division of the Department of Finance.[67] Even in their efforts to help maintain Inuit lifestyle, Canadian officials thus drew Inuits into the welfare, tax, and other bureaucracies of southern Canada.

The Canadian Arctic supports only a modest population of hunter-gatherers who have followed the seasonal movements of animals across the vast landscape for millennia. Baffin Island, at 500,000 square kilometers, the fifth largest island in the world, supports a population of under twelve thousand people in scattered communities across the Arctic desert. Such mammals as caribou feed on moss and lichen. In the absence of agriculture, the hunters relied entirely on the caribou, following them across the landscape. Their sleds and dogs were indispensable to life. In the absence even of driftwood, they built frames for shelters out of whale bones and covered them with sod or with the skins of caribou and seals, and they used walrus hide for their boats and sealskin for their kayaks.[68]

The number of dogs belonging to an Inuit family depended on the productivity of the hunting territory and the skill of the individual hunter. Such technological change as access to guns, nets, and steel harpoons increased the ability to hunt and led families to keep more dogs. The rise of intensive white fox trapping and a trade economy triggered another rise in team size to between ten and fifteen animals. Yet the Inuit dog almost disappeared with the introduction of snowmobiles because of their speed, their reach, and their simplicity: it takes only minimal maintenance to keep a snowmobile going, whereas it takes many hours to develop a good sled dog team. The snowmobile also permitted weekend hunting.[69] Paul St. Onge observes that snowmobiles and other technologies changed traditional habits of mobility among the Inuit in Kangiqsualujjuaq (Nunavik) in a way that led to the westernization of their society. While it is impossible to separate the influence of snowmobiles from that of other technologies, snowmobiles were the crucial piece in the consolidation of the "western" way of living because they gave the Inuit "the technology needed to practice traditional activities in a contemporary context."[70]

After considering decades of evidence on the detrimental effect of technological change on Inuit hunting communities, Greenland's lawmakers banned the snowmobile for hunting. Ever since motorized boats first appeared in Greenland in the 1920s, the Inuit have understood that the sounds and smells of engines scare wildlife away. According to Frank Sejersen, a professor of Eskimology at the University of Copenhagen in Denmark, the Inuit sensed that engine pollution and other factors have forced beluga whales to migrate further out to sea, "which has made hunting more difficult." Speedboats and rifles together led to the decimation of entire bird colonies. Sejersen observes, "The delicate balance of the polar ecosystem has been tipped in favor of overwhelming human predominance by hunting technology." The snowmobile pushed the balance further. Said Aqaluq Lynge, president of the Greenland Inuit Circumpolar Conference, an Inuit advocacy group, "Snowmobiles will destroy the hunting grounds, and what is the point of traveling this way? You can't see anything move around you. You don't appreciate anything." In Greenland, the solution was to ban the snowmobile.[71]

The same pattern of rapid social change and decimation of fragile ecosystems has repeated itself in the Svalbard Archipelago in the Norwegian high Arctic,[72] in Canada, in Alaska, and in the lower forty-eight states. Towns and entire regions have become dependent on snowmobiles. Farmers and city dwellers alike have changed their lifestyles to accommodate their machines, spending thousands of dollars to get away on the weekends, in the process unwittingly destroying unique ecosystems.

Snowmobile Safety and Public Health

Snowmobiles offer both uproarious fun and great potential for serious injury or even death. The operation of snowmobiles—a high-speed vehicle intended for operation on ice and snow—carries risk. Medical personnel have been aware of the dangers of snowmobile operation for decades, although public recognition of the dangers and steps to improve safety have been slower in coming. One of the first reviews of this danger appeared in 1970 in the *Journal of Trauma*.[73] The authors analyzed data from northern New England, including records from the Mary Hitchcock Memorial Hospital in Hanover, New Hampshire, and questionnaires returned by twenty-four area physicians. Regional doctors obtained data on fifty-nine persons involved in snowmobile accidents in the 1968–69 season, some of whom had been partially or totally paralyzed. On the one hand, the authors observed that mushrooming use of snowmobiles had made the gathering of statistics difficult. The absence of a "uniform system for reporting snowmobile accidents" may have led to undercounting. Yet the evidence showed a growing crisis, with "legislative control of their operation [failing] to keep pace with the soaring popularity of the snowmobile." Anyone, with or without training and experience, could legally drive a vehicle almost anywhere, at any time, without regard to speed or weather conditions, and certainly without lights. The doctors worried that regulators had adopted a wait-and-see approach, as a result permitting accidents to increase in frequency, and only later would they "somehow or sometime get around to recognizing the problem and plan countermeasures."[74]

Specialists outside of North America also noted a sudden increase in accidents. In Jämtland, a county of northern Sweden with about 6,600 snowmobiles in use in the early 1970s, 117 people were injured in accidents, a rate of 1 per 55 snowmobiles.[75] As the number of snowmobiles registered in Sweden increased, so did the number of accidents, and a standard epidemiology came to be associated with them: drivers tended to be male, to have been using alcohol, and to have been driving too fast. Accidents were most likely to occur at night. Many drivers drowned when they and their vehicles plunged through thin ice on lakes and rivers.[76] By the early twenty-first century, snowmobile fatalities in Sweden had become a national concern, with physicians calling for educational programs to fight high-speed operation and driving under the influence, and also to encourage the use of snowmobile suits and helmets.[77] In Finnmark County of northern Norway, injury statistics revealed a higher accident rate for snowmobiles than for road traffic.[78]

An eighteen-year review of snowmobile injuries in northern Newfoundland and Labrador indicated a similar epidemic in Canada. Snowmobiles had by the mid-1980s become widely popular in the region as a mode of transportation, recreation, and sporting life. At the same time, the number of accidents treated annually at one hospital in St. Anthony, Newfoundland, had increased sixfold between 1969 and 1986. The authors of the study called for enforcement of legislation, intensified public education about the hazards of snowmobiles, and modified engine design to provide increased protection for the lower limbs.[79]

As with ATVs and personal watercraft, emergency room personnel and law enforcement officials grasped the growing epidemic of snowmobile accidents long before there was any move toward developing a public safety program. Without collection and analysis of data on a national level, and without the requirement that state and local data be gathered systematically, public health officials remained woefully uninformed. In the 1990s a series of reports published in medical journals and by the Consumer Product Safety Commission and the Centers for Disease Control (CDC) changed the situation. In 1991, surgeons at the University of North Dakota School of Medicine reported on eighty-eight patients injured over ten years in snowmobile accidents. Most of the patients were male, and seventeen (19.3%) of them were under age sixteen. Inexperience, alcohol use, carelessness, failure to follow manufacturers' recommendations, and excessive speed contributed to the accidents.[80] At the same time, physicians recognized that snowmobile injuries were often extensive and required costly hospitalization to treat. In one study published in 1996, researchers found that for forty-two individuals who suffered snowmobile-related bone fractures, blunt abdominal trauma, head injury, and other injuries, hospital costs averaged $16,227.[81] Within a decade, those costs would triple. A CDC mortality and morbidity weekly report in 1997 indicated that death rates had increased state by state, and that the death rate per vehicle was also increasing.[82]

Deaths and serious injuries hit all age groups in every state, although young men, usually encouraged by alcohol, are the ones most likely to have a snowmobile accident. In New Hampshire, twenty-six deaths occurred from 1982 to 1992. All operators involved in the fatal accidents (13) and most involved in nonfatal accidents (161 of 188, or 86%) were young men. No operator involved in a fatal accident and only 7 percent of those involved in nonfatal accidents had taken an off-road vehicle safety course. Inclement weather was not usually a factor, but darkness was. Operating on a frozen body of water was a factor in five of the nine fatal accidents and in one-quarter of the nonfatal accidents. Alcohol and excessive speed also played a role.[83]

No state was immune. During the 1995–96 winter season, both the Maine Department of Inland Fisheries and Wildlife and the Maine Office of the Chief Medical Examiner detected an increase in deaths associated with snowmobile use. From the fall of 1991 through the spring of 1995, three to eight snowmobile-related deaths occurred each winter season. In the 1995–96 winter season, twelve people died. While the number of registered snowmobiles increased from 61,641 to a record 76,477 between 1991 and 1996, the death rate per registered vehicle in 1996 was higher than in any of the previous five years. Of the thirty-nine deaths, thirty-two (82%) resulted from trauma and seven from drowning; thirty-seven (95%) decedents were male. Twenty-five of them (81%) were wearing helmets at the time of the incident, and fifteen (41%) were legally intoxicated. The accidents usually occurred during clear weather (79%). Excessive speed often played a central role (52%). In an effort to fight this epidemic, Maine state law required snowmobile operators (residents and nonresidents) to report all incidents involving snowmobiles that result in injury requiring medical attention or in property damage amounting to $300 or more. But they neither required helmet use nor restricted use among children.[84]

By the twenty-first century, snowmobile accidents had become epidemic. More than two million people operated the machines on a regular basis. But how many of them could control the 600-pound bullet that reached speeds of 90 mph? What did these machines have in common with those developed to help move people and supplies for utilitarian purposes? The fact that snowmobile accidents produced roughly two hundred deaths and fourteen thousand injuries annually indicated that the new machines had little in common with the first generation. Physicians clamored for regulation and legislation.[85] Canadian doctors were no less adamant in their call for serious measures to make snowmobiling safer. By 2002, 16 percent of severe sports and recreational injuries were caused by snowmobile accidents, a much higher rate than for such winter sports as skiing and snowboarding.[86]

Children were often the unintended victims, a phenomenon especially noticeable in rural communities. Several trends were clear: males younger than sixteen were three times more likely than females of the same age to sustain a snowmobile-related injury. During 1995 there were 16,226 snowmobile-related injuries, with one-fifth occurring to persons younger than sixteen; from 1993 through 1995 there were 10,628 injuries to children with a total cost of $84 million, or $8,000 per injury. Evidence indicated that children lost control with greater frequency than other groups, lacked adequate skills to operate a snowmobile, were often not properly trained, and were too small to control the weight, speed, and power of a snowmobile.[87]

The significant mortality and morbidity among children led the American Academy of Pediatrics to determine that "recreational operation of snowmobiles is inappropriate for children and adolescents." Academy members urged snowmobilers to travel at safe speeds, especially on unfamiliar or rugged terrain, avoid the use of alcohol or other drugs before or during the operation of a snowmobile, wear appropriate clothing, carry a survival kit and cell phone, travel in groups, pay attention to weather to avoid hypothermia and frostbite, avoid ice, not carry more than one passenger, and keep their vehicles well maintained. Then the Canadian Pediatric Society—though not manufacturers, legislators, regulators, or proud parents—urged that off-road vehicles should be banned for use by children under fourteen years of age.[88]

Another persistent problem was the fact that snowmobiles, like their mud and water relatives ATVs and jet skis, were increasingly subject to product-safety recalls. In the early 1980s the problem attracted the attention of regulators. Following reports of serious injuries involving snowmobile drive tracks, Kawasaki agreed to a $3 million to $4 million voluntary program to repair approximately 16,500 1978- and 1979-model snowmobiles. In May 1980, officials at the Consumer Product Safety Commission (CPSC) learned of the problem with the snowmobile tracks, which by the time of the recall had led to at least twenty injury incidents (fractures, severe lacerations, and in three cases injuries requiring the amputation of a finger). Kawasaki also agreed to pay $90,000 to settle its alleged failure to report the danger promptly to the CPSC.[89] Between 2000 and 2005, snowmobile manufacturers recalled over fifty thousand vehicles.

Undeniably, snowmobile manufacturers produce first-rate vehicles, and virtually all mass-produced vehicles have at some time experienced a recall in the name of product and consumer safety. But the recalls, along with other problems associated with these machines—excessive noise, pollution, irresponsible use—have on occasion given snowmobiles a bad name. A network of local clubs works to address these problems head on.

Snowmobile Clubs: In Defense of Access to Wilderness

Recognizing that they cannot defend noisy, highly polluting snowmobiles and wishing to avoid heavy-handed government interference in their production, purchase, and use, the American Council of Snowmobile Associations and the International Snowmobile Racing Specialty Manufacturers Distributors Group have publicly stated their opposition "to excessive sound levels that result in restrictions against snowmobiles." They believe that few other factors contribute more to "misunderstanding and prejudice" against the snowmobiling community than excessively noisy machines. Industry representatives claim that a minority of operators give a bad reputation to the rest by gunning engines, and they point out that snowmobiles are built to federally mandated noise control standards. (They do not mention the possibility that those standards may be weak.) Still, they recognize that ignoring noise will likely result in "excessively rigorous state and federal standards, more expensive and less attractive snowmobiles, the reduction of choices in aftermarket products, abusive enforcement of current laws and other solutions undesirable to riders and the snowmobile industry."[90] Hence the trade associations recommend that riders be sensitive to community standards. They urge organizers of snowmobile events to use advertising, peer pressure, and enforcement to make excessively loud machines unwelcome. They advise retailers to discourage the installation of loud replacement exhaust systems, understand that modification doesn't lead to improved performance, and educate riders that excessive noise contributes to fatigue with potential impact on their enjoyment and skills.[91]

As to what constitutes "excessive" noise, the trade associations respect individual choice but urge operators to consider the time of day, the traffic mix, the surroundings, and people nearby. The Council and Racing Group issued a statement of its concerns in response to pressure to close trails in the United States, knowing that in Europe road closures to stifle excessive noise have become commonplace, and anti-tampering legislation and restrictive sound emission requirements may follow. If not addressed "voluntarily, and in a timely fashion, these restrictions are inevitable." And "the right to ride a snowmobile does not permit us to infringe on the peaceful enjoyment of life by others. Indeed, many others, including the courts, view snowmobiling not as a right but a privilege."[92]

The International Snowmobile Manufacturers Association admits that snowmobiles are noisy. Pre-1969 machines emitted sounds as high as 102 dB at 50 feet. By 1972, states began to curb snowmobile noise, setting decibel limits on noise heard at 50 feet at full throttle. Noise levels have been reduced 94 percent since then, and especially since 1975, when they had to meet the Snowmobile Safety and Certification Committee standard of 78 dB at 50 feet when traveling at full throttle, and 73 dB at 50 feet while traveling at 15 mph.[93] Yet snowmobiles, Reich reports, are noisy for technical, economic, and social reasons: the two-stroke engines are cheap, powerful, but loud, and manufacturers avoid using baffled mufflers, which disrupt exhaust flow and rob the machines of power. Many sportsmen also simply prefer noisy machines.[94]

More could have been done with existing technology. By the early 1970s, several manufacturers employed water-cooled engines. The cooling system was either an automobile-type radiator or a tubing network in the engine compartment and under the seat. Temperatures in liquid-cooled engines can be stabilized, equalized, and lowered, eliminating hot spots and extending spark plug life. And "because the engine doesn't have to have an opening to the outside air, it can be completely enclosed by sound insulation in the shroud, and a water pump uses perhaps half the power of a fan." Air-cooled engines had the advantages of never freezing, boiling, or corroding and operated at higher, more efficient temperatures.[95] But they require fins to cool, and the fins amplify sound. Industry representatives claimed that trying to meet some sound standard—say, a limit of 80 dB at 50 feet for snowmobiles on public lands—would change the machine itself so much that it would no longer be a snowmobile.[96]

Another issue is whether snowmobiles ought to meet higher standards for emission control. The U.S. Environmental Protection Agency has proposed new emission regulations for snowmobiles that would take effect in 2006. The EPA noted that, as of 2002, snowmobiles emitted 220,000 tons of hydrocarbons and 580,000 tons of carbon monoxide annually, emissions that also included benzene and carcinogenic particulates. The new standards would not require retrofitting but would apply only to new snowmobiles built in 2006 or later. Manufacturers would be permitted to meet the standards on a fleet-average basis, meaning that some engines could be cleaner and some dirtier than the standard. EPA officials noted that the standards would raise prices by $50 for a two-stroke engine, $300 for a direct-injection engine, and $900 for a four-stroke engine but that fuel and other costs would be significantly lower, so that the improvements would pay for themselves over the life of the machine.[97]

The manufacturers opposed the regulations but have had no

trouble meeting them. For example, Yamaha's RX-1 and RX-1 Mountain four-strokes met the new standard in 2003. They get 30 percent better fuel mileage than other machines. According to Yamaha, "Both motorcycle and sledding enthusiasts love the distinctive exhaust note of a four-stroke, and it's quieter than traditional snowmobile engines." Honda announced that its Gold Wing 1800 and Interceptor motorcycles "exceed the stringent 2008 CARB (California Air Resources Board) emissions requirements years ahead of schedule."[98] The new Honda four-stroke jet skis meet both CARB 2004 and EPA 2006 emission standards. The result is cleaner air and quiet for everyone to enjoy out of doors.[99]

Snowmobile clubs, the central institution to promote responsible use, have grown in popularity along with snowmobiles themselves. Some state clubs have more than twenty thousand members. Snowmobilers are drawn together by their love of their sport, the desire to promote responsible snowmobiling, and what they feel is a pressing need to protect snowmobiles and snowmobile trails from restrictions, encroachment, and regulation. Much more than other recreational machine clubs, snowmobile clubs tend to be supported primarily by such local commercial businesses as restaurants, motels, outfitters, gas stations, and general stores. This may be because they are so vital to local economies on a seasonal basis. National organizations claim a total $38 billion contribution to the nationwide economy of snowmobiling. Local clubs tend to get less funding than they need from states and communities to help maintain trails, so they need business sponsors to supplement annual membership dues.

Clubs tend to be strong advocates of law-abiding snowmobiling precisely because they recognize that this will protect their access to land. They understand that their sport is a privilege rather than a right. They hold ATV operators responsible for much of the bad press they receive. They claim to respect the environment, although many of them seem not to comprehend how their sport intrudes upon nonmechanized recreationists and wildlife. They are family-oriented, sponsoring special "women's days," activities that cater to children, and so on. They abide by a "snowmobile code of ethics" that was developed by a committee of representatives from the U.S. Forest Service, Bureau of Outdoor Recreation; the Michigan Department of National Resources; the Minnesota Department of Natural Resources; the Department of Lands and Forests, Ontario, Canada; the U.S. National Park Service; and snowmobile manufacturers. The code is a model of clarity and common sense:

1. I will be a good sport. I recognize that people judge all snowmobile owners by my actions. I will use my influence with other snowmobile owners to promote proper sports conduct.
2. I will not litter any trails. I will not pollute streams or lakes.
3. I will not damage living trees, shrubs, or other natural features.
4. I will respect other people's property and rights.
5. I will lend a helping hand when I see someone in distress.
6. I will make myself and my vehicle available to assist search and rescue parties.
7. I will not interfere with or harass hikers, skiers, snowshoed hikers or other winter sports enthusiasts. I will respect their rights to enjoy our recreation facilities.
8. I will learn and obey all federal, state and local rules regulating the operation of snowmobiles in areas where I use my vehicle. I will inform public officials, as required, when using public lands.
9. I will not harass wildlife. I will avoid areas posted for the protection or feeding of wildlife.
10. I will only use marked trails, areas or roads open to snowmobiles. I will not travel cross-country when prohibited.

Snowmobile club activities indicate the essential community focus of members. In 1990 the Roseau City (Minnesota) Trailblazers snowmobile club drove to Falcon Lake, Manitoba, 200 miles to the north, as part of a $3,000 fundraiser for trail maintenance. One by one, roughly a hundred snowmobilers passed through the customs inspection. In 1999 an effort to establish a Guinness world record for one thousand snowmobiles setting off one after the other for an international trail ride failed when "only" five hundred showed up because it was so cold. The line of snowmobilers still reached 16 miles.[100] The Trailblazers club had withered away in the late 1960s and early 1970s, but by 2005 there were 500 miles of trails, 300 in Roseau County alone. "The club also polices itself. It educates its members on where to go and where not to go."[101]

The Association of Wisconsin Snowmobile Clubs (AWSC), one of the largest such organizations in the United States, consists of over six hundred clubs throughout the state totaling thirty thousand family members. Its council consists of one representative and one director from each of seventy-one Wisconsin counties. These directors and representatives keep up to date on legal issues facing snowmobiling "and in general work together to keep Wisconsin snowmobiling safe. The directors/Representatives then pass that information to their counties, and from the counties the information then goes to the clubs."[102]

The AWSC sponsors an annual "Miss Snowflake" competition open to young women who participate in snowmobiling. In partnership with the Liberty Mutual Insurance Company, the club also offers scholarships to applications who submit the best "what snowmobiling means to me" essay. In 2004–5, Ms. Samantha Booth was chosen to represent the AWSC as Miss Snowflake. She was both honored and excited by that award. She wrote, "I am looking forward to attending many events through out the year and meeting many of you. My parents (Jeff and Sue), brother (Mitchell), and I belong to the Wheeler Snowdrifters of Dunn County. We live on a small hobby farm outside of Wheeler. I am currently attending the University of Wisconsin [at] River Falls, which is where I reside Monday through Thursday. Aside from school and snowmobiling, I enjoy spending time with friends and family, hunting, fishing, ice skating, swimming, baking, jogging, working with our animals, tractor pulling, and many other outdoor activities."[103] In an email to me, Miss Snowflake shared in greater detail the reasons she feels honored to have won the award and how great a role the snowmobile played in her life. She wrote:

> I began snowmobiling at the age of five on a child size snowmobile and have been an avid rider ever since. I try to do as much riding as possible with friends and family when we have snow. Many

> times I pack up with friends and family and travel to the snow and put on some miles. . . . Snowmobiling is truly a family sport. My family and I belong to the Wheeler Snowdrifter's Snowmobile Club. . . . Many clubs stay active in that respect by hosting get togethers/social events. Some people in our club don't even own snowmobiles, but they enjoy the work and the time the club spends together. . . . [Everyone] is really involved and responsible for protecting snowmobiling in Wisconsin. As Miss Snowflake for 2004–05 I travel through out the state of Wisconsin and promote snowmobiling as the safe family sport that it is. I attend many club functions—fundraisers, snowmobile safety courses, social events, etc. I also write articles for the Wisconsin Snowmobile News. . . . All in all, snowmobiling is a huge part of my life. Snowmobiling has taught me many things. It has allowed me to be more independent as a young rider and has shaped me into a responsible adult. It truly is a wonderful family sport.[104]

Miss Snowflake was a superb choice as an ambassador for her sport.

Snowmobile club members host banquets, races, and the like to raise funds to help keep the trails in the excellent shape they are in and to purchase new grooming equipment. To become a member of a Wisconsin snowmobile club, you pay $20 (for a family), and if your club is a member of the AWSC (most are), you receive the *Wisconsin Snowmobile News Magazine.* According to Miss Snowflake:

> Snowmobile Clubs in Wisconsin are in charge of marking, maintaining, and grooming their section of the trail. They also work with private property owners to get permission to have a trail cross their land. Most clubs show their appreciation to private property owners by hosting a banquet for them or by giving a gift certificate. Truly, without the Clubs and the private property owners there wouldn't be any trails to ride on. It is these people along with the Association of Wisconsin Snowmobile Clubs (AWSC) that make the trails possible. . . . These members are strictly volunteers, they donated their time for the love of snowmobiling as well as the camaraderie between members.

Miss Snowflake also explained how safety and education go hand in hand:

> To operate a snowmobile all by yourself on the trail you must be at least 12 years old and have obtained your snowmobile safety certificate. This holds true for everyone born on or after January 1, 1985. People older than this cut off date are not required to have obtained a snowmobile safety certificate, however it is valuable knowledge to have and is strongly recommended. I did actually ride my snowmobile into school when I was in high school.[105]

Age limits, helmet requirements, and other safety regulations differ from state to state, with many state regulators not displaying nearly enough common sense as public health, safety, and environmental concerns would seem to warrant. Generally, states require snowmobiles to be registered. Each state has somewhat different laws about access to roads; most prohibit road use except for short stretches to facilitate access from one trail to another. Snowmobilers in most states have secured exemptions from laws requiring helmets. Only Michigan, Nebraska, New York, Pennsylvania, and Vermont require helmets for all users. Minnesota, New Hampshire, North Dakota, and Utah require helmets only for those up to the age of seventeen or eighteen.[106]

Supporters of recreational vehicles often point to their positive impact on local economies in order to justify efforts to maintain access to parks in competition with other nonmechanized recreational uses. Snowmobile clubs and local and state governments have commissioned economic impact studies to determine just how extensively and in what ways recreational machines generate revenues. In August 2003 two economists at Plymouth State University produced a study for the New Hampshire Snowmobile Association that indicated direct and indirect spending of approximately $666 million during the December 2002 through April 2003 snowmobile season. They also calculated a total impact on the state's economy of nearly $1.2 billion. These calculations grew out of a survey of snowmobilers that gauged their travel, machine use, lodging, eating, and other practices, then extrapolated the results to the some sixty-nine thousand snowmobiles registered in the state. The researchers determined that the expenditures directly supported 65,557 full- and part-time jobs, although not in any high-tech industry.[107]

A study in New York determined that during the 2002–3 season, snowmobilers in that state had an economic impact of up to $875 million, compared with $476 million in a 1998 study. Again, the researchers undertook a survey of 5,000 of the 166,000 registered snowmobiles in the state; they received 1,361 responses, on the basis of which they calculated the profile of the average snowmobiler (age, sex, income, spending habits) and ascertained attitudes toward clubs, regulations, and licensing fees. While nearly 90 percent of users in a 1998 survey were male, by 2003 only 60 percent were male. Snowmobilers were, not surprisingly, wealthier than average citizens, judging by household income. Most users owned snowmobiles, not rented them. More than half of the snowmobilers belonged to a club, but fewer than one-third had taken a safety course. Most of them supported the idea of raising fees so long as the funds were used to pay for the extension and maintenance of trails.[108]

Similarly, the Wyoming Recreation Commission, in conjunction with the University of Wyoming, prepared a report that showed that snowmobiling was responsible for $189.5 million in economic impact in the state in 1998, while the University of Maine and the Maine Snowmobile Association demonstrated an economic impact of $261 million.[109] Miss Snowflake drove home the importance of the machines to local economies:

> Many small businesses (restaurants, hotels, gas stations, etc.) thrive off the winter income from the snowmobilers. The counties in Wisconsin produce a county map of the trails. On this map they sell advertisement space to these businesses and encourage snowmobilers to support these sponsors. The AWSC also sells ads for the Wisconsin Snowmobile News Magazine (Commercial Sponsors on the web site) and also encourages snowmobilers to show their support year round.[110]

Of course, one could say that any growing sector of the economy generates jobs. Waste management generates jobs. But should communities have the right to prevent snowmobiles or incinerators from congregating in them anyway? Does the presence of a snowmobile make recreation for nonmotorized users difficult? The snowmobile economic stimulus studies do not address such questions. Nor do they consider or subtract the environmental costs of snowmobiling, the social costs of deaths, injuries, and lost workdays, and the law enforcement costs.

Snowmobile organizations have avoided systematic consideration of the impact on public health and the environment. Instead, in the face of efforts to limit or ban them from a number of sites, they have responded with a vitriolic defense of the rights of snowmobilers to go into parks over the rights of others not to share the blue smoke and high-pitched whines. Ed Klim, of the International Snowmobile Manufacturers Association, consistently puts the debate over how best to manage public lands in terms of light and dark, freedom and tyranny, truth and bias, science and emotion, democracy and elitism. He asserts that such organizations as the Wilderness Society represent elitist views while the manufacturers speak for the ordinary American. In an essay called "'Wilderness' Means 'Keep Out,'" he accuses opponents of unrestricted machine use of creating "spin" (which he himself apparently does not create) in their efforts to control the "nation's landmass . . . by their rules and under government control." The elites speak in "righteous tones" and use "scare tactics to raise money for their coffers." With more than 100 million acres of public land already closed to machine recreation (an area roughly the size of Maine, New Hampshire, Vermont, Massachusetts, New York, and Pennsylvania combined), the Wilderness Society, Klim asserts, seeks to place even more land off limits, lobbying for another 200 million acres to be designated as wilderness by 2010.[111]

Let us recall the concerns of Wilderness Society founder Robert Marshall. Marshall cautioned strongly against allowing machines to dictate policy. He had watched government building programs for visitors' centers and roads expand through national parks and forests since the 1910s. He understood that the federal government had promoted highway building in the national parks in the 1930s as a way to fight unemployment. But roads and wilderness were incongruent, and the plan to build more of them to permit automobiles to clog the woods was an abomination. "The automobilists argue that a wilderness domain precludes the huge majority of recreation-seekers from deriving any amusement from it," he wrote. "This is almost as irrational as contending that because more people enjoy bathing than art exhibits we should change our picture galleries into swimming pools." There were already more roads than all the automobiles in America could cover in a lifetime. Roads would "molest the few remaining vestiges of the primeval."[112]

Nevertheless, the all-terrain vehicle, off-road vehicle, and snowmobile followed the automobile into the woods, leading to unending devastation of the pristine world. Today, some 130,000 miles of official trails cover the nation (see table A.3), and who knows how many tens of thousands of miles of illegal trails. Funding for the trails comes from snowmobile registration fees, gas tax rebates, trail permits, and volunteer construction and maintenance, supplemented by grant-in-aid programs to spread and maintain trails not only on public but on private lands. Trail design, according to manufacturers, takes into consideration safety, the environment, destination, and the integration of the network—the latter to such an extent that "it is now possible to ride 70% of North America's snowmobile trails through . . . continuous interconnections especially in the eastern part of the continent." If 70 percent of the trails are connected, it means that ecosystems have become fragmented.[113]

Incongruous Wilderness: The Democratic Process and Technology

Snowmobiling is an important weekend sport for the millions of American families who support the activity through clubs. Yet many analysts argue that small but wealthy public interest groups have increasingly gained access to crucial public health and environmental issues to the detriment of the sport, as they have inordinate power over the policy process generally. Critics of this excessive power, on the right and on the left, note the flood of funds into election campaigns and into political action committees against the best efforts to regulate or limit them. They refer to the central position of lobbyists, not only in pushing legislation but in drafting it. A narrower case of the way in which powerful yet unrepresentative groups have manipulated the democratic process concerns the place of technology—including its promotion and its regulation for safety and efficacy—in twentieth-century America.

In the 1960s and 1970s, several bureaucracies, old and new, gained new responsibilities to regulate various technologies and processes in the interests of public health, workplace safety, and environmental preservation, or to provide advice on those technologies and processes. These include the Environmental Protection Agency, the Food and Drug Administration, the Nuclear Regulatory Commission, and the short-lived but effective Congressional Office of Technology Assessment. Extensive legislation supported the effort so that government departments and bureaucracies as well as private organizations and businesses were required to conduct far-reaching evaluations of the costs and benefits of new technologies ranging from drugs and foods to power stations and automobiles. In many instances an environmental impact statement was required, and publication of intended rule making in the *Federal Register* became a recognized feature of the regulatory process. The process enabled concerned parties from citizens to public interest groups to business groups and manufacturers to comment on regulations. The National Environmental Protection Act, the Clean Air and Clean Water acts, new statutory authority granted to such organizations as the Office of Safety and Health Administration and the Consumer Product Safety Commission, and a series of other laws and enabling acts unquestionably improved the quality of air and water, led to the clean-up of hazardous waste, and improved the safety of consumers and workers.

Beginning in the 1980s and continuing with greater energy to this day, antigovernment activists, industry representatives, property rights proponents, and others raised concerns that regulation had gone too far. Some of them argued that government regulation amounted to interference with appropriate private sector business activities. Others worried that the federal government had usurped not only states' rights but also those of individual property owners in telling them what they could or could not do with their land. Businesses were concerned that regulation would require them to retool or otherwise modify production and thereby place onerous costs on them to meet unjustified standards. How exact all of these observations may be is subject to some argument. But the case of snowmobiles (and ATVs and personal watercraft) indicates that the pendulum has swung to the side of antiregulation activists, so much so that there is a real danger that wilderness areas and national parks and other federal lands may suffer irreversible damage. The ruination will come either through the increasing use of lands solely for the short-term economic value of timber, minerals, oil, grasslands, and other commodities, with substantial scars left behind, or through a misguided effort to permit machines to coexist with other recreational uses.

The ability of snowmobile manufacturers to introduce models that meet more rigorous safety, noise, and pollution standards versus claims that the new standards will impose undue and costly burdens on the industry; the nature of the public health costs of snowmobiling; the extent and permanence of its effects on ecosystems and wildlife; and above all the rights of snowmobilers, including those very dedicated, responsible, family-oriented club members, to secure access to trails versus the rights of other recreationists to hike without machine intervention—all of these issues have come to a head in the controversy about whether to permit snowmobiles in national parks. In that light, the George W. Bush administration's decision to roll back efforts to limit access of snowmobiles in Yellowstone National Park in Wyoming and Denali National Forest in Alaska signifies a misguided effort to privilege the machine at the expense of all other uses. The dispute over snowmobile access to Yellowstone reflects the larger debates over what constitutes wilderness, what a national park is and how it should be used, the extent of snowmobiles' environmental impact, and the degree to which they should be regulated to ensure the preservation of lands important to all Americans, operators and nonoperators alike.

The National Park Service allows snowmobiles in forty-three units (11%) of its parks, seashores, monuments, and other lands in apparent violation of Nixon and Carter executive orders. Outside of Alaska, where snowmobiling is permitted by law in parks, the most popular national park for snowmobiling is Yellowstone, which had seventy-six thousand snowmobile visits in the 1999–2000 season. In January 1998 the Bluewater Network and sixty other environmental organizations petitioned the park service to ban snowmobiles in compliance with executive orders and in view of their impact on air and water quality, public health, and safety. Bluewater officials pointed to growing evidence of the harmful effect of snowmobiles on such endangered and threatened species as Canada lynx, grizzly bear, bald eagle, and gray wolf. They referred to data showing how packed snowmobile trails changed predator-prey relationships. They noted noise levels commonly exceeding 100 dB.[114] The Clinton administration followed through in upholding the law by ordering the EPA and park managers to produce an environmental impact statement. The statement showed that the costs to the park and to park rangers of maintaining snowmobile access were too high. In keeping with the interests of a vast majority of citizens and to protect the parks from irreversible damage, the government ordered a ban on snowmobiles. But the Bush administration backed away from this ban and intends to allow the vehicles to enter the parks after all, presumably until such time as global warming has melted all the snow.

Members of Congress have addressed the matter of snowmobiles in national parks primarily through provisions of appropriations bills that establish winter use rules for Yellowstone and Grand Teton National Parks. Disputes over the appropriations generally reflect a dispute between those who wish to phase out snowmobile use, who are largely Democrats, and those whose interests more closely resemble those of the snowmobile industry, who are largely Republicans. The Bush administration rule would set a daily limit of 950 snowmobiles for Yellowstone, 115 in Grand Teton, and 400 on Rockefeller Parkway connecting the two. On most days the limit would result in no reduction from current levels and perhaps an increase; on weekends and holidays it might result in a reduction.[115]

Opponents of snowmobile access argue that the resulting pollution harms park rangers as well as wildlife and detracts from the wilderness experience at Yellowstone. "Employees inhale gasoline fumes while performing their required duties and Park Rangers have long complained of experiencing nausea, headaches, dizziness, and hearing loss when working near snowmobiles," Public Employees for Environmental Responsibility observed in a lawsuit. Others refer to Yellowstone's "special quality" in winter. "It is a rest time for the park or at least it was before the onslaught of the snowmobiles," said Marcus Libkind, director of the Snowlands Network, a group that promotes "human-powered" winter activities. "It should be the one time of the year when the park is as close as possible to being free of motor vehicles. It is a time when wildlife is struggling to survive and we, intruders on their ground, should do as little as possible to add stress to their lives." Libkind also notes that even though modern four-stroke snowmobile engines are relatively clean and quiet, they still transfer urban noise levels to the wilderness. The noise of a city or "of an interstate highway should not be the basis for Yellowstone. The same goes for the smell of motor vehicles."[116]

Snowmobiles came to Yellowstone not because of user demand—after all, there are millions of acres on other lands nearby for snowmobilers to use—but because of business interests that sought winter activities to sustain restaurants, hotels, and other economic concerns. Over the years tourism has brought thousands upon thousands of visitors into the national park, turning what had been a winter wonderland into a noisy thoroughfare filled with noxious blue smoke. If snow coaches seating eight to twelve persons were used instead of individual snowmobiles, winter use could go on with significantly lower impact. But industry and business see short-term profit in Yellowstone, although overuse will certainly lead to ecological degradation and ruin the park's profit-making attraction. At each stage of the controversy, as more and more snowmobiles have entered the park and nonusers have sought to limit those numbers, officials have conducted studies to examine the impact. They always conclude on the basis of scientific investigation that significant costs to persons and the environment occur that outweigh the benefits to a small if vocal group of businesspeople and snowmobilers.[117]

Michael Yochim argues that a seemingly innocuous decision decades ago not to plow Yellowstone roads opened the way for snowmobiles into the park. Administrators thought initially that the snow machines would be safer and more convenient than automobiles and would not require the Park Service to expend its limited annual funds on road plowing. Suddenly, it was impossible to keep machines out of Yellowstone, especially near Old Faithful. Park superintendent John McLaughlin concluded in the park's 1965 annual report, "It seems inevitable [that] mechanized over-the-snow travel may replace skis and snowshoes. . . . Undoubtedly more Park travel during the winter months by this type of machine can be expected and should be encouraged. This type of recreation is increasing rapidly in this particular section of the country. . . . [The] machines are now relatively inexpensive and maintenance requirements simple. Much of the terrain of the Park and its features are compatible and attractive to this mode of winter travel."[118] Over the next three decades the number of snowmobile visits grew from hundreds to thousands to tens of thousands of visits annually.

Park administrators' actions accelerated park use in the 1970s. So did the opening of two nearby ski resorts: the multi-million-dollar facilities at Jackson Hole, about 50 miles south of Yellowstone, and the Big Sky Ski Resort on the West Fork of the Gallatin, about 30 miles north of Yellowstone. The park itself permitted concessionaires to expand the Snowshoe Lodge Facility near Old Faithful and open a hotel at Mammoth Hot Springs. They permitted all sorts of activities that have little to do with nature and much more to do with amusement parks, and especially with money-making: snow coach tours, snowmobile rentals, cross-country ski rentals, sleigh rides, hot-tub rentals. Employee dorms in several instances were transformed into hotels. The park itself opened "warming huts" with wood stoves and fast-food services.[119] Welcome to McYellowstone National Park.

By the early 1980s, over seventy snowmobile-related businesses stood at Yellowstone's west entrance. Dean Nelson, president of First Security Bank, acknowledged that "the winter economy is the snowmobile." West Yellowstone billed itself as the "snowmobile capital of the world." In recognition of the efforts of successive superintendents to facilitate entry of machines into the park, the International Snowmobile Industry Association, which later sued the park service for attempting to protect it from machines, awarded the service the association's International Award of Merit—for "enlightened leadership and sincere dedication to the improvement and advancement of snowmobiling in the U.S."[120]

This "enlightened leadership" has contributed to the destruction of Yellowstone. Snowmobiles generate 27 percent of all carbon monoxide emissions and 77 percent of the hydrocarbons annually, even though they represent a very small percentage of the vehicles that use the park—and do so for only four months a year. During the 1992–93 season seventy-seven thousand snowmobilers visited Yellowstone. Annual visits have since averaged sixty thousand. They leave a blue pall of smoke; the Old Faithful site often smells of unburned fuel. Under pressure to allow sixteen-, twelve-, and even eight-year-olds to use snowmobiles, administrators allowed not only more machines but more operators into Yellowstone: twelve- to sixteen-year-olds were permitted to operate a snowmobile when supervised by parent or guardian (within 50 yards).[121]

The Clinton administration determined, on the basis of clear evidence of damage to the parks, to their wildlife, and to personnel, that the only rational scientific policy was to ban snowmobiles. The Bush administration has reversed that effort with a phase-in of cleaner snowmobiles and a shift of snowmobile traffic to other parts away from the main entrance at West Yellowstone and Old Faithful. Under the Bush plan, no more than eleven hundred snowmobiles a day will be allowed in Yellowstone, in neighboring

Grand Teton National Park, and on the parkway connecting them. Only half that number would be allowed to pass through West Yellowstone, the most popular entrance for snowmobilers, on any given day. Since the parks have had an average of 840 snowmobiles daily in winter but up to 1,650 a day during holiday and other busy weekends for the past decade, the new rules would allow more snowmobiles into the parks on average while cutting numbers only on the busiest days.

Recognizing the danger to national parks of permitting continued noise, pollution, and degradation, a number of environmental groups sued to require the tougher standards that can be achieved with current technology. The federal courts agreed with them. Ms. Vickie Patton, a lawyer with Environmental Defense, a plaintiff along with Bluewater Network, said the court understood that the EPA must put in place tough standards to protect public health and the environment from snowmobiles. "The court's decision shows that there is no place for these high-polluting engines when cost-effective and clean air solutions are at hand." On top of this, the Bush EPA twice rolled back already weak emission standards proposed for off-road vehicles including snowmobiles, even though current four-stroke snowmobile engines had achieved substantially greater emission reductions than the new standards required.[122]

Ed Klim contends, however, that the Bush administration rules were "extremely stringent" and that it was unrealistic to require manufacturers to change standards without more lead time.[123] Yet can anyone in the United States, the home of space shuttles and jet engines, of microwaves and computers, of an unrivaled automotive industry, doubt that snowmobile manufacturers are incapable of building appropriate engines? Those engines are already available, and surely this important industry, with $1 billion in annual sales, with 110,000 units sold in 2003–4, and with fifty years of experience, can build safer and cleaner machines to meet the standards, not to mention its own rhetoric of responsible use.

Seven "stewards of America's national parks" who had served collectively nine presidents wrote Secretary of Interior Gale Norton to protest vigorously against the intention of the Bush administration to permit snowmobiles to continue ruining Yellowstone National Park. They included George Hartzog, National Park Service director from 1964 through 1972; Denis Galvin, its deputy director during 1985–89 and 1998–2002; and Michael Finley, Yellowstone superintendent from 1994 to 2001. They pointed out that all environmental studies of the impact of snowmobiles in Yellowstone, including a supplemental study commissioned by the Bush administration, had concluded that phasing out snowmobile use "best preserves the unique historic, cultural and natural resources associated with Yellowstone and Grand Teton National Parks." Ignoring the reports' conclusions, they wrote, would "clearly be to accept avoidable risks to health and safety [and] a narrower range of beneficial uses." They pointed to public comment in favor of the phase-out that ran four to one against maintaining snowmobile access. The study concluded that keeping the snowmobiles would cost taxpayers $1.3 million more each year than replacing snowmobiles with snow coaches.[124] In the meantime, the levels of glucocorticoid in elk and wolf blood and skat continue to rise.

As the Bush administration made clear its intention to ignore the best scientific studies and maintain access to Yellowstone for special interests, a bipartisan group of congressmen introduced an amendment to the Interior Appropriations Act in summer 2003 to phase out snowmobile use. Led by Rush Holt (D-NJ) and Christopher Shays (R-CT), they sought through the Yellowstone Protection Act to protect Yellowstone and Grand Teton parks. The act was not aimed at users or the industry, since users and industry have so many other places to encourage snowmobiles to cavort. In discussion of an amendment that would have prohibited funds being used to manage recreational snowmobile use in the parks, Holt declared that Yellowstone, the symbol of America, was "being loved to death."[125]

Congressman Joseph Hoeffel (D-PA) argued that "government policy is virtually to require the use of snowmobiles. . . . They do not pave the roads so that cars can ride into Old Faithful or around Yellowstone [but] they groom the roads with snow on it and pack it down for the use of snowmobiles." The technological solution of four-stroke engines, Hoeffel pointed out, did not solve the problem. They too polluted and were noisy. Rangers had to wear respirators to combat the noxious smoke of hundreds of snowmobiles. Supporters of the amendment noted that the regional economic impact of snowmobile phase-out was exaggerated. In 1995–96, when visits to the West Yellowstone entrance decreased by more than 13 percent in connection with controls on snowmobiles, resort tax collections increased by 10 percent. This meant that preserving Yellowstone against snowmobiles attracted more visitors and raised more revenues than did preserving snowmobile access. And such local papers as the *Great Falls Tribune*, the *Helena Independence Record*, and the *Caspar Star Tribune* sided with the phase-out, for the Bush administration alternative would be "a dirty, stinking shame." The amendment failed as Republican members of Congress voted overwhelmingly against it.[126]

On October 14, 2003, Holt wrote directly to Interior Secretary Gale Norton seeking clarification as to why the department was determined to maintain snowmobile access to the parks in defiance of the environmental impact statements and adverse public commentary. Norton's decisions in this matter, Holt wrote, were a "derogation of the Secretary's and agency's responsibilities" to protect park employees, wildlife, and natural resources.[127] Responsible snowmobilers, those family farmers from Minnesota, Wisconsin, and elsewhere who created these wonderful modern machines, also deplore Norton's dereliction of duty.

The Glories of Snowmobiling

Author and range manager emeritus Thad Box laments the changes to his favorite spots wrought by air-cooled small-bore engines. In the Cache National Forest, Box finds refuge from roads and machines in a grove of mixed conifers and aspen beside a small stream. It's a world in which "terrorists, traffic, and famine give way to bluebells and columbine beside clear water gurgling over rocks." At first, he shared the spot "with only cows, deer, elk, moose, chipmunks, and hummingbirds." Later his wife, and then their children and grandchildren, joined him. A few old fire circles indicated other human presence, but only rarely did they encounter a hunter. Then snowmobilers "forced their machines through an old cow trail that winds up the creek. Tree branches were broken. Soil eroded where vegetation was destroyed on a 'shortcut' across a sagebrush hill." They grew more aggressive and destructive, churning up soil and uncovering roots of fir trees, leaving them susceptible to disease. The snowmobilers "played tag in the little opening near the creek. Moss-covered stream banks were mashed into mud. Streamlets cut new paths where machines chewed through the meadow. Beds of fir needles were scattered. Tracks dug through melting spring snow and plowed-under wildflowers." Box considered this damage a significant loss to humanity, caused by insensitive people simply having "fun." He noted that "fines or putting people in jail will not bring back loved ones or magic spots." Yet "punishing those who cause loss in pursuit of their own pleasure or ego enhancement is not as effective as educating folks to think about the results of their actions." Box argues, "Associations, whether recreationists or livestock growers, need to police themselves. Presidents of snowmobile associations in Utah and Idaho tell me they are increasing their educational programs. A dedicated snowmobiler now carries a digital camera on his machine. If he sees someone abusing the land, he photographs the action. He sends a picture to his association. If that doesn't work, he sends the picture to the Forest Service."[128]

"We sell snowmobiles for joy," Mitchell Johnson of Polaris told me. The snowmobile permits people to be in the "pristine" outdoors, where they can follow their "primeval" instincts. "There's a thrill in speed, in being able to go places in an unfettered way," Johnson said. "My father taught me that he built the snowmobile to enjoy the wilderness." Ultimately, he believes, there is "a huge legitimacy to ATVs, snowmobiles and watercraft" for the utility, camaraderie, and exhilaration they provide.[129]

Like all responsible machine recreationists, Mitchell Johnson recognizes that there must be limits to machine use, and like them he believes that machine use is compatible with environmentalism:

"We have the Boundary Waters Wilderness Area. I love the wilderness. There's no motorized traffic allowed there. It's wonderful and I'm glad." But some areas "must be designated multiple use. It must be controlled but if a snowmobile facilitates seeing Yellowstone in the winter, good." Mitchell continued, "I'm not a radical environmentalist. I have an SUV to pull my snowmobile." The question is, "How do we enjoy, create, use these products and still preserve this world in a way that will sustain itself. How can we as a company be stewards of our sport."[130] That is indeed the question.

In the effort to maintain a modicum of health in several Rocky Mountain ecosystems inhabited simultaneously by cattlemen, mining consortia, and farmers, state and federal officials have sought to reintroduce species of once-numerous megafauna, such as wolves. Unlike wolf populations in southern Canada and the contiguous United States, which have nearly been eliminated, those in Alaska and northern Canada remain abundant. Yet there too the wolf has come under pressure, in large part because of machines. According to Marco Musiani and Paul Paquet, some biologists "are concerned about the killing of wolves in certain areas of northern Canada. They consider such commercial hunts a problem because of the vulnerability of individual wolves to the specific hunting techniques employed, particularly the use of snowmobiles." Public opinion too has begun to turn against using snowmobiles to hunt wolves, on ethical grounds. There is no challenge to killing wolves from a snowmobile: "Hunters can use snowmobiles to quickly approach escaping wolves until they are within range of a rifle."[131] A utilitarian taking of nature has given way to an unthinking taking of nature by hunters mounted on the back of snowmobiles.

The paradox of the snowmobile is that in its fifty years of existence it has undergone significant innovations in suspensions, steering, engines, comfort, and safety and yet remains noisy, dangerous, and highly polluting. Some safety improvements, such as night lights, better braking systems, and dead-man throttles, came about under pressure from government regulators, while most others grew out of the vision of technical excellence imagined by the engineers at Polaris, Arctic Cat, and Bombardier. Snowmobiles permit exploration of wilderness by solitary individuals, yet they also destroy that wilderness, threatening wildlife and denying all recreationists the solace of nature. Unless manufacturers and users themselves demand safer machines and more responsible use, another paradox will arise: significant limitations on the use of machines designed to take you almost anywhere on cold, snow-driven, socked-in, isolated days.

❮ 3 ❯

INTERNAL COMBUSTION ADVENTURE

ATVs ARE THREE- or four-wheeled motorcycles. Massive, knobby, low-pressure tires provide buoyancy as they move across nearly any terrain, their powerful engines lifting them up and down hills, through mud and streams, along sand dunes. ATV club members—and there are many, many clubs and many, many club members—are by and large a law-abiding group. ATVers are family-oriented, friendly people; they are your next-door neighbors. They use ATVs to commune with nature. They respect private property, they clean up their litter, they wear helmets, they know the limits of their machines, they stay out of emergency rooms. They carry winches, an optional purchase, to pull their 600-pound vehicles free when they get stuck, a frequent occurrence. They hope to have a tree, a rock, or another ATV nearby to hook on to. Given the terrain and conditions under which ATVers operate, the need for a winch is understandable.

ATV manufacturers have built marvelous machines, improving stability and engine performance over the years and mass producing them in sizes, function, and horsepower to meet any consumer demand. They have found a place on the farm, on the ranch, at the utility company, in the logging industry, in the military. With attachments they plow, furrow, grade, haul, lift, and remove snow. ATVs were originally manufactured in three-wheeled versions. Manufacturers stopped producing three-wheelers at the end of the 1980s in a consent agreement with the federal government because they were prone to tip and cause severe injury, although they may still be sold and are available for purchase on the private market. Also by voluntary agreement with federal regulators, but not required by law, four-wheeled ATVs can be sold to families for use by children, but with no bigger than 50-cc engines for six- to twelve-year-olds and no bigger than 90-cc engines for twelve- to sixteen-year-olds. ATV manufacturers believe that these restrictions, established by working with federal and state officials, are sufficient to ensure rider safety while enabling the operator to enjoy his or her machine. They—and owners—believe that if you're foolish enough to do something foolish on an ATV, then that's the breaks. A Honda ATV safety brochure lists its registered trademark safety slogan as "Stupid Hurts."

ATV enthusiasts and manufacturers have secured tax, license, and other fees to subsidize the maintenance of trails. They maintain that only irresponsible owners stray from designated areas and trails, and that you should not penalize all riders, no more than you would penalize all automobile drivers, for the illegal activities of a few. And if 5,000 Americans have died using ATVs and 1.4 million have been hospitalized, this is usually because of irresponsible operators, not the machine. Wrote one ATV supporter, "The true crime here was that ATVs were deemed harmless by riders: they didn't have motorcycles' tendency to fall over immediately when ridden by unskilled operators. Most of the public-relations damage was inflicted by uneducated, non-helmeted, beer-guzzling adults riding like total goons and unsupervised kids crashing adult-only machines."[1]

When the federal and state governments turned to rules and regulations to force manufacturers to make their vehicles safer, an action required because of loss of life and injuries that accumulated with ATV use during the 1970s and 1980s, the manufacturers began to argue in such trade publications as *ATV Connection* that too much regulation would "dummy down" the sport. Greg Hall, the technical editor of *ATV Connection,* contended that manufacturers were being forced to build ATVs for the lowest common denominator, something so foolproof that any person who could walk and chew gum at the same time could ride one. This was lawyer-based engineering geared to protecting uncoordinated people from themselves, not market-driven, cutting-edge engineering.[2]

The result is today's modern, powerful, fun-to-operate, utilitarian machine, although far more people use ATVs for sport, hunting, touring, and recreation than for utility. ATVs have massive size and footprint. They weigh in at over a quarter-ton, and their 48-inch width enables them to go virtually anywhere. Huge 150-hp engines power them up and down treacherous terrain. Operators love to take them through streams and gullies, along dunes, inevitably widening paths. Five major manufacturers—Honda, Yamaha, Bombardier, Polaris, and Suzuki—now produce roughly one hundred thousand annually. There are seven million so-called utility, sport utility, and sport model ATVs. Their owners trail them everywhere, every weekend.

ATVs have reached public consciousness on three levels that indicate an ongoing reevaluation of their place in American culture. The first concerns the huge number of enthusiasts who worry about restrictions on where they can ride and who oppose requiring such safety systems as roll bars and seatbelts. The second concerns the public health epidemic that has resulted from ATVs. The annual cost of medical care, according to the American Academy of Pediatrics, is $6 billion. The third centers on the extensive damage they have wrought on the environment in ecosystems from fields to forest and from wetlands and riparian ecosystems to deserts and other arid climes. Children have paid the highest cost with their lives, and those who survive ATV accidents in their youths will discover less wilderness in which to spin their tires. The ATV is both a beauty and a beast. *ATV Scene*'s monthly Miss ATV, draped alluringly over a powerful machine, does little to temper the danger. Unlike Miss ATV, Ms. Jeanna Darnell, a contestant in the 2004 Miss Texas Pageant—a beauty representing Rio Grande Valley, a singer, twirler, and pianist, the owner of over 150 Russian *matryoshka* nesting dolls—sees the dangers of these machines clearly. At pageants her platform presentation is a speech on "ATV Safety—ATVs Are Not a Toy."[3]

Many people believe that ATVs require redesign to protect public health and the environment and must face greater restrictions on their use on such public lands as parks, monuments, wilderness, and other sensitive areas. Too many ATV users trespass on public and private property, convincing critics that they do not understand trail limits. They have established 500,000 miles of legal backcountry roads and trails already, and no one knows how many hundreds of thousands of miles of illegal trails. Trails exist in wilderness areas, in corridors near protected areas that suffer accordingly, and in once-quiet neighborhoods. Also, like other recreational vehicles, ATVs pollute. After being prodded to do so by federal action, manufacturers have begun to produce cleaner and more efficient engines, although for decades to come, highly polluting one- and two-cylinder two-stroke engines will remain the norm. ATV engines also produce a high-pitched 96-dB noise. At an average of 350 miles traveled per ATV per year—a total of 23 billion miles, although the Environmental Protection Agency estimates that the mileage is in fact much higher—these machines damage soils, degrade water, kill wildlife, pollute the air, spread invasive species, and contribute to aesthetic loss.[4]

Several interrelated social, economic, and political trends contributed to the rise of ATVs and other off-road vehicles (ORVs). First, the managers of state and federal lands have long promoted motorized recreation and have been slow to recognize its evident costs. In the 1920s they supported the construction of roads, visitors' centers, and concession stands in parks in response to the automobile boom. In the 1930s, as part of the national effort to recover from the Great Depression, they put laborers to work building more roads through parks. As for forests and other lands, land managers built roads there too as part of their charge to ensure fair use of federal lands for forestry, mining, and recreation. In the postwar years the newfound wealth of Americans enabled more of them to seek recreation via automobile and boat in parks and forests. Gasoline was plentiful and cheap. The ATVs took to the roads and trails. Finally, the ATV, essentially a four-wheeled motorcycle, was relatively simple to mass produce on the basis of existing industrial capacity, with only modest retooling. The ATV has become a fixture of the forests, the woods, the beach, and suburbia—all this from modest agricultural roots.

The Pinto of Outdoor Recreation

The impetus for the first off-road vehicles including the predecessors of ATVs came from people who loved mud. *Mechanix Illustrated, Popular Mechanics,* and *Hot Rod* publicized these vehicles, which resembled hotrods but quickly became utility vehicles. One such ORV was the Coot, which handled mud, mountains, lakes, and woods with ease, enabling the farmer or outdoorsman to "hunt, fish, mend fences, find stranded sheep and haul fertilizer." Forest rangers found the Coot useful for safety and law enforcement purposes. Sales of the vehicle grew to eighteen hundred in 1968. The Coot celebrated versatility; it carried four passengers or 1,000 pounds at 25 mph, could handle grades up to 75 percent, and could cruise through water at 1.5 mph. The power plant was a 12-hp lawn mower engine.[5] By the late 1960s at least fourteen American companies sold ORVs in the price range of $1,400 to $1,800. The editors of *Mechanix Illustrated* encouraged readers to build their own vehicles, publishing specifications on various fun buggies that could be built from metal tubing, plywood, and an 8-, 10-, or 12-hp engine.[6]

One early mass-produced option was the Moto Brousse, turned out in the thousands in the 1960s and 1970s. *Mechanix Illustrated* called this vehicle "lightweight" at 450 pounds. The front wheels were larger than the rear. The vehicle didn't sink into soft surfaces, and its steel body was perfect for "busting through heavy brush or splashing through streams," yet narrow enough to go where a wider ATV could not.[7] Another early model produced by Standard Engineering in Ford Dodge, Iowa, was an amphibious, center-articulated ATV that could climb grades of up to 45 percent and came with a propeller to move at 4 mph in water. Without the propeller, the 12-hp Eagle could grade roads, plow snow, or cut the lawn.[8] The name "muckmobile" left no doubt about its purpose.[9] Manufacturers saw unlimited possibilities for these vehicles—for example, for camping, and specifically for bringing the civilized world along with you when you camped. The ATV Manufacturing Company, of Glenshaw, Pennsylvania, produced a trailer as a way to avoid "extremes" when camping. Rather than having to "carry your shelter on your back into the rough country or drive off in a largish vehicle that may resemble a motel on wheels," the camping trailer carried all camping needs and even floated.[10] As engines grew more powerful and suspensions more versatile, operators encountered fewer and fewer obstacles to off-trail activity.

ATV racing geared to adults and children followed. Shock-absorbent balloon tires with heavy treads were made for race courses indoor and out, artificial and natural, over and through anywhere: hills, rocks, sand, shallow water, and mud. Wrote a historian of the sport, "ATV racers enjoy challenges. They race side by side on narrow tracks. . . . They drive through mud pits and over obstacles. They slide through turns at high speeds. ATV racing is very dangerous." The dirt tracks featured sharp curves, hills, jumps, and irregular terrain. But "almost anyone can race ATVs. Both adults and children race ATVs. Boys and girls can start racing when they are six years old. All racers must wear safety gear and know safety rules before they can race." The manufacturers stopped making racing ATVs as too expensive, even as the number of national events topped eleven hundred annually with up to nineteen different classes competing. Contemporary racing ATVs are therefore modified stock vehicles.[11] As for other recreational machines, the racing circuit was important to generate customer loyalty and to encourage purchase of insignia items. All companies sponsor teams, some having half a dozen motorcycle and ATV riders. Polaris's team includes eleven-year-old Jerry Welsh from Oakland, Maryland, a Grand National Cross Country circuit competitor who has been racing ATVs since he was six years old. The companies often sponsor race series, for example Suzuki with the Grand National Cross Country circuit. There are at least seven different race series including the ITP Quadcross, Extreme Dirt Track ATVA, World Off Road Championship Series, ATV Cross Country Series, and ATV Desert Baja Series.

The first real ATV sold in the United States was the Honda 90, a relatively simple motorcycle technology that lent itself to mass production. It gained popularity among Japanese farmers for its versatility on the farm when used with a trailer or other attachments. It could carry tools, manure, crops, and small animals. Honda executives realized that consumerist America provided a great market for this simple machine, but they did not initially understand just how much Americans would love them for leisure rather than work. True, Honda owners in America found dozens of uses for their ATVs as utility items, at home and at work. Being less expensive and lighter and more maneuverable to operate than pickups or tractors, and having a remarkably light footprint (with wide low-pressure tires), they could cover almost any terrain and were used in ranching, industry, police work, and farming. But recreation has always been the major use. Honda engineer Osamu Takeuchi developed two-, three-, four-, five- and even six-wheeled configurations. Three wheels seemed to work best. Honda engineers used a 70-cc four-stroke single-cylinder engine with an extended rear axle for the ATV (hence the ST 70). They developed 2.2-psi tires based on AmphibCat tires. They discovered that a 90-cc engine was needed to push the amphibian tires (hence the Honda 90). In 1970 the first U.S. 90s appeared, selling for $595. The ATC 90, as it became known, became the ATC 110 in 1979. Honda engineers developed new tires that were less vulnerable to punctures.[12]

During the 1980s the popularity of the ATV grew rapidly. Utility and thrills were the two reasons. Regarding utility, the significantly higher cost to purchase, maintain, and run a standard farm tractor brought ATVs to farms. But approximately 80 percent of Honda's market in the 1980s was for multipurpose uses (utility *and* recreation). In 1980 Honda introduced the ATC 185, which had large 25-inch tires, a five-speed transmission, an automatic clutch, and a 180-cc four-stroke one-cylinder engine. The ATC 250R, introduced in 1981, was a high-performance ATC with a 248-cc air-cooled two-stroke engine, adjustable suspension, front disc brakes, and five-speed transmission. Suzuki built its first four-wheel ATV in 1983. Kawasaki, Yamaha, Honda, and Polaris soon followed suit. Honda's first four-wheel ATV, the TRX 200, debuted in 1984. The market responded rapidly: in 1984 Honda sold 370,000 units, 69 percent of the U.S. total.[13]

Over the years, manufacturers added various improvements: bigger, more powerful engines, automatic transmission, electric starters, disc brakes, better shocks and suspension, and safety improvements to combat the tendency to roll over. Improvements in air-cooled engines followed. The high point for the ATV may have been the 650-pound Grizzly 600 four-by-four, introduced in 1998, with a 595-cc engine (versus 2.2 liters in my 1990 Honda Accord), dual front disc brakes (which my 1999 Caravan lacks), and ultramatic transmission. ATVs generally come with 400-cc or larger single- or twin-cylinder air- or liquid-cooled four-stroke engines, independent MacPherson strut–type suspension, rack and pinion steering, and disc or four-wheel sealed hydraulic drum brakes. Utility ATVs weigh half a ton or more and carry payloads weighing twice that. They have sealed beam lights. Adult sport ATVs weigh 350 to 600 pounds but come with the same size engines as utility ATVs. In some ways these are automobiles rather than ATVs. My Accord and its 2.2-liter engine weighed in at 2,733 pounds, so the ratio between weight and engine displacement in ATVs and automobiles is roughly the same, while horsepower per weight is significantly higher in ATVs and gas mileage lower.

Bombardier, a world leader in powerful machinery from subway cars to logging equipment and recreation vehicles, offers a wide range of ATVs. These include the Outlander, Rally, Traxter, DS and DX series, and Youth series. The Outlanders are sport utility vehicles with Rotax four-stroke liquid-cooled engines that range from 400 cc to 800 cc, have automatic transmission, full instrumentation, disc brakes, MacPherson strut or double a-arm front suspension, and weigh between 600 and 700 pounds. They can pull half a ton or more. The seat is 3 feet off the ground, making for a high center of gravity. The vehicles, Bombardier tells us, are "a potent mixture of horsepower, torque and control." The Rally is a bottom-of-the-line Outlander. Outlanders seem to have replaced Traxters. The DS series are sport ATVs geared to thrills and hopefully not spills. They range from the DS650X, which is "the most powerful production big bore sport ATV," to the DS250, which has a 249-cc engine. The Bombardier slogan, "Follow No One," encourages riders to establish new trails. The manufacturer's advertisements advise, "If you're not the lead dog, the view never changes." Bombardier riders "are not just thrill seekers, but thrill finders." Another admonishment is to "Ride all day. Never touch the ground."[14]

Polaris has done well with its ATVs. Its Sportsman line, the best selling in the industry, complemented the trail-riding Predator 50, the "utilitarian" All-Terrain Pickup, the mid-sized Phoenix for young adults and women, and three youth models. The special Predator 500 Troy Lee Edition was designed in collaboration with a California firm of the same name, and the Sportsman 800 Twin EFI came with a more powerful engine and dual exhaust, making it "the biggest, baddest ATV ever."[15] In 2005 *ATV Sports Magazine* named the 500 TLD Predator the "Sport Quad of the Year." This was the second time in three years the Polaris Predator had earned the award. In its three years on the market the Predator has won seven "Sport ATV of the Year" awards from a number of industry publications. The 2005 TLD had new gear ratios to improve acceleration, aluminum shocks with compression adjustability, and Maxxis Razr radial tires. Jerrod Kelley, editor of *ATV Sports Magazine,* wrote that the Predator "does not alienate the average rider or recreational rider, yet it has the potential to perform in competition formats." To achieve the award a machine must be all new or significantly redesigned. It must also meet at least one of three prerequisites: market appeal, competition potential, and innovation.[16]

ATVs quickly filled dealerships' lots. They were fun, easy to operate, and a good way to separate Americans from their disposable income. ATV sales have grown fourfold since 1990, with six hundred thousand units sold in 2000 alone. ATVs represent one of the fastest-growing forms of outdoor recreation: riding over uneven terrain, through mud and streams, on a powerful, high-speed vehicle that can weigh three times more than the operator in its saddle.

ATV manufacturers have chosen model names for their vehicles that indicate a fascination with power over nature: Yamaha's Big Bear, Wolverine, Warrior, Kodiak, and Raptor; Polaris's Explorer, Trailblazer, and Magnum; Bombardier's Outlander, Traxter, and Quest; Suzuki's King Quad, Quad Racer, and Quad Sport. Some models are amphibious. All are intended to give the owner the sense that he is pushing the frontier, exploring and trailblazing where no man has gone before—or so he might think, if not for the ruts and erosion he encountered.

ATV manufacturers have received Department of Defense funding to underwrite their development of powerful machines for the civilian sector. In 2004 the Department of Defense gave $10 million to Polaris for military-modified ATVs that "inspire new models in the civilian market" such as the Sportsman MV—"like those used by US forces in Iraq."[17] The Sportsman MV truly looks military, like a cross between a jeep and an ATV, painted in camouflage green. It has a liquid-cooled twin-cylinder 683-cc four-stroke engine, independent long-travel rear suspension, on-demand all-wheel drive, electronically activated front and rear 2,500-pound winches, heavy-duty flat racks, oversized D-rings with a capacity of 450 pounds, steel-reinforced frames, racks, and floorboards, automatic transmissions, MacPherson strut front suspensions, and four-wheel disc brakes. The military connection in ATVs suggests how incongruous ATV use and environmentalism are. The Department of Defense has long sought to gain authority from the U.S. Congress to ignore environmental protection laws on its bases in the name of national defense, in part to try out new machines for use in war theaters.[18]

One with Nature on an ATV

Because they are not restricted by climate to winter and by geography to the snowy northern states, ATV clubs are more widespread than snowmobile clubs. The lack of geographic and climatic restrictions may have led these clubs to be less formal; riders can get together at almost any time in almost any weather for a muddy trek through the woods. ATV club Internet sites reflect this informality. They have forums where ATVers discuss their sport, make plans to meet up to ride, troubleshoot repairs, and talk about the next purchase. Many ATV clubs have no officers or board of directors, for example the Mud and Dirt Club in Alabama. In the west, federal officials assist ATV enthusiasts in finding places to ride. In the east, state governments and clubs are crucial to securing weekend playgrounds. In both regions, clubs are crucial to maintaining access to lands, establishing trail systems, and opposing what members believe to be unfair and unnecessary regulation.

ATV and other recreational machine clubs are nonprofit groups. The IRS recognizes them as tax exempt under section 501c7 of the code as organizations that serve educational functions. Clubs encourage their members to ride safely, to follow state and local laws, to "pack it in, pack it out," and so on. Yet in their support of the business of ATVs sales and service, club members are not strictly speaking engaged in educational activities, since both directly and indirectly they help manufacturers to sell machines, clothing, insignia items, and the like. Still, clubs are an important force in getting members to observe the applicable laws, including those involving registration, fees, and taxes, and to avoid excessive resource use. ATVers have established clubs for every type of machine, all terrains, and a wide range of activities. Clubs represent all walks of life, both genders, and every color, race, and creed, although membership reflects ownership: most owners are white middle-class males, and there are few clubs for gays and lesbians or people of color. All clubs have as their goals bonding in the experience of machine recreation and preserving access to trails, parks, and bodies of water. Many of them seek to promote family values as variously defined in the different clubs. They encourage responsible use, although a small number of them exist specifically for hellacious riding. The most organized clubs diligently follow regulatory efforts to keep members up to date about current legislative initiatives that might affect them.

The members of the Central Oregon Motorcycle and ATV Club (COMAC) frame their activities as a way of addressing the problem of young people using drugs or getting in trouble with the law. "Remember a family that plays together stays together," they insist. The club has 600 miles of trail system created to counter what members saw as efforts to regulate them out of existence. "The club was started in 1988 by a group of families that saw what was happening to the off-road riding in Central Oregon and around our nation. We [off-road enthusiasts] were getting beaten up in the news and on television, and most importantly, our riding areas were being closed. Since then, COMAC members have worked hard to make Central Oregon a great place to ride and recreate." COMAC works closely with other clubs, the Forest Service, and the Bureau of Land Management to help maintain public land, educate riders, and make the most of their sport. Their motto: "We All Have One Thing in Common—Our Love of Riding and Our Commitment to Keep Our Sport Alive and Well in Central Oregon."[19]

ATV clubs stress that they are environmentally conscious, but the reality is more complex. Their mottoes often indicate a different mind set, for example "Mud, Ruts and Guts" for one club. The Bamaboggers ATV club sets out to keep areas open for all. The photos that accompany their website indicate that they intend to have good muddy fun, and that everywhere they go there will be mud. They are interested in "going deeper, farther and faster than anyone has gone before from the deepest swamps to the highest mountains . . . in Alabama, Mississippi . . . and many other states." Along with mud and ruts comes patriotism. Clubbers claim great allegiance to America, a country whose forefathers and foremothers, they maintain, guaranteed their rights to the wilderness. The members specifically connect the protection of their rights to drive anywhere, and the struggle to maintain access to lands, with past American struggles for freedom. Riding for them is a right, not a privilege.[20]

Riding is considered such a right that most states do not require ATV users to buy liability insurance for property damage or personal injury. One exception is New York, whose Department of Motor Vehicles website states that "you may not operate any ATV anywhere in New York State, except on your own property, unless it is covered by liability insurance. Minimum required coverage is $50,000/$100,000 for death, $25,000/$50,000 for injury and $10,000 for property damage in any one accident." Pennsylvania may be the only other state requiring liability insurance. If ATV operators or bystanders are injured in an ATV accident and they also lack health insurance—likely a frequent occurrence, given that over forty-six million Americans have no health insurance—then the burden of their treatment when they come to emergency rooms with fractures, trauma, and internal bleeding falls on taxpayers.

Operators can afford, however, automatic transmissions, quick bike twist throttles, reverse lights, tows, and winches. They can afford specialty tires with trade names like "Mud Bug" and "Mud Shark." They gladly purchase such optional equipment as dual and single compound grips for hand bars, nerf bars (stainless steel tubes affixed to the ATV like running boards), graphics kits, storage and cargo racks, queen seats, tube bumpers, foot pegs, helmets, boots, clothing, and snowplows. Club members know that ATVs lose value quickly, with $5,000 vehicles often reselling for under $1,500. In addition to the machine itself, operators usually spend another $1,000 to outfit themselves and the vehicle. Manufacturers recommend wearing a helmet and goggles at a cost of $100 to $125, and good models at $250, although many states do not require adults to wear helmets. Boots run another $150, with gloves, rain suit, jerseys, and pants adding $250 to $350. The accessories for the machine include a front fuel rack that carries two 1.5-gallon poly fuel tanks ($125), lounger seats for the back with storage compartments ($250; these come with the admonishment "Not for passenger use," which ATV passengers universally ignore), cooler racks ($65), and fender bags ($24). For those operators with more wanderlust, distributors recommend a trailer cart ($330), chrome exhaust ($220–$270), and wireless remote winch ($420) to pull the ATV out of a ravine. Don't forget the GBC Dirt Devil tires ($60 each), or perhaps the Sand Tire Unlimited Mud Machine tires ($95) or Maxxi's Mudzilla tires ($85–$110). The trailer will run another $2,000. For the utility sportsman, another option is the 900TR fully hydraulic grapple loading trailer, complete with Honda 5.5-hp power unit for logging purposes ($8,800).

Members of ATV clubs see no conflict between their activities, their love of nation, their love of nature, and what ATVs may do to fragile ecosystems, the animals that live in them, and the people who dislike their noise and pollution. They equate their love of ATV activities with patriotism. According to the Ozark Trails ATV Club:

> Welcome to the beautiful world of OHV's [off-highway vehicles] and the people who ride them! We are a varied group of individuals held together by a common love of and respect for this great land in which we had the privilege of being born and raised. We don't ask much of it, only the right to enjoy it in our own way, just as those who are younger and stronger do in theirs. We ride trails and back roads, and obey the rules and regulations. We are not here to harm the environment but we are here. We volunteer many hours to help care for our public lands. We know they are not ours alone, but, just like the streets and highways, they belong to us all. We leave only tracks where we pass, and we are willing to pay our fair share of the cost of maintaining our God given heritage. We didn't ask to be born Americans, but we were, and we support the freedoms fought for and won by our magnificent warriors both past and present. So, don't count us out, and don't underestimate us, for we are strong, and we have the determination to fight for the right to ride! If you agree with us, come and ride with us. If not, tell us where we are wrong. Let's get together and work for the good of us all![21]

The 4wheelers4Christ see ATVing as the way to stay close to Jesus. They dedicate themselves to "Uniting Christian 4Wheelers World-Wide" by providing a place to congregate, both on the Internet and in the woods, for individuals, clubs, families, and businesses "united in a common objective—to spread the word of Jesus Christ in the 4Wheeling community and beyond." As one reverend four-wheeler wrote:

> The rush of climbing an awesome hill, the sounds of the tires finding their grip over jagged rocks and through slippery gullies, the thrill of reaching the top, the simple pleasure of seeing a part of nature that very few people get to see, and the bond that comes from being with friends who enjoy a similar passion! Yes, off-roading provides something that the rest of our lives doesn't touch, and sends the kind of inner message that says, "Life doesn't get any better than this. I love being out here." . . . Remember, taking your 4×4 off-roading is one of the greatest outdoor pleasures there is, but it is nothing in comparison to having a personal relationship with the God of the universe. No matter how great the weekend, you know at some point you have to come back to reality. When Jesus lives in your heart, the adventure never ends![22]

State officials across the nation offer commonsense suggestions concerning club organization, officers, potential meeting places, bylaws, meetings, and so on. Some of the advice inadvertently promotes what ATVs actually do. Regarding club names, for example, some officials urge members to "avoid names that landowners may find offensive or suggest improper use such as 'Mud Slingers.'" Since the evidence indicates that one pass by an ATV can irreparably damage such vital ecosystems as stream beds and wetlands, one way to limit further environmental degradation is to utilize existing cleared areas, logging roads, abandoned roadways or railroad rights-of-way and other recreational trails. The officials instruct users to avoid bodies of water, tree plantations, sensitive wildlife areas, and areas with precipitous terrain, and they urge club members always to gain permission of landowners before use.[23]

ATV clubs derive strength from their numbers. With the help of manufacturers, business sponsors, and state agencies they secure access to the out-of-doors and the funds to maintain and expand a network of trails. One of the largest such organizations, the New York State Off-Highway Recreational Vehicle Association (NYSORVA), acts as a conduit for communication with such state agencies as the Department of Environmental Conservation for a steadily growing user community estimated at well over three hundred thousand riders and their machines. Members seek "increased[,] safe and environmentally-conscious OHV recreational opportunities" on both public and private lands in New York.[24] Every time a state government allocates funds for trails, it takes funds away from another program. Each trail draws more riders. And each new rider requires new trails. How have state governments responded?

States, Municipalities, and ATVs

Since the early nineteenth century, state and local governments have sought to promote technologies for commercial purposes. They have encouraged the construction of roads, bridges, railroads, and highways in pursuit of commerce. Legislators, councilmen, and businesspeople have pushed for canals and dams to improve transport, control flooding, and facilitate irrigation. Entrepreneurs have sought sympathetic ears in state legislatures to gain tax breaks, rights of way, or outright grants of land to build industries and create jobs. So the active role states have taken in supporting recreational machines is not surprising, although their reluctance to acknowledge the dangers associated with these vehicles by restricting access to certain areas or requiring universal helmet use certainly is. The involvement of government officials seems innocuous at first, because elected officials and their staffs should support publicly approved activities.

Generally speaking, in western states with extensive open spaces and access to federal lands, legislators have been much more hesitant to regulate the use of recreational machines. They see restrictions on access to parks, forests, and the trails through them as a violation of states' rights. They often reject the designation of lands as monuments or wilderness areas even though that designation is intended to protect them from overwhelming human assault. They believe that individual choice should determine helmet use. Governments in eastern states tend toward greater restrictions and greater concern about individual safety. They have had no choice, given the tighter spaces, the fewer miles of trails, and the therefore much more visible, immediate impact of riders and their machines.

Efforts to regulate ATVS in New York, Maine, and New Hampshire illustrate the difficulties involved in securing reasonable laws and policies. Like other state and federal officials, those at the New York Department of Environmental Conservation (DEC) face great challenges in establishing an equitable balance among the various aspects (trails, education, enforcement, and stewardship) of ATV administration. Only late in the game did New York officials recognize the need for ATV regulation, licensing, and liability laws. In 2000 and 2001, reacting to pressure from hikers and conservation groups, the department closed several public areas to ATVs. The department then commenced more active enforcement of recently passed regulations pertaining to Adirondack Park, where illegal and inappropriate use of ATVs on state forest preserve lands had become rampant. Rangers informed the Forest Preserve Advisory Committee that the sale and use of ATVs in the Adirondacks and west to Syracuse were "skyrocketing." Illegal and careless use was "out of control." Catching illegal ATV users on forest preserve land was "difficult in the best of times, but more so when patrol times for Forest Rangers are cut due to increasing administrative responsibilities." And tickets to operators for trespassing or for violating vehicular or conservation laws did little to stop them. Towns in the Adirondacks had passed ordinances regarding ATVs that sent operators deeper into the woods, traveling without respect for ownership boundaries. Towns that permitted ATVs on their roads encouraged more trespassing on private lands and illegal entry onto state lands than did towns that prohibited or seasonally regulated ATV use.[25] The Adirondacks were not alone: illegal ATV use has spread to almost every park in every state.

Under mounting pressure from all sides, DEC officials resolved to consider the place of ATVs from environmental, revenue, and licensing points of view for all state lands.[26] In March 2005 the DEC acting commissioner, Denise M. Sheehan, released a draft commissioner's policy for public ATV access to recreational programs on state-owned land.[27] The proposed policy was intended to ensure that ATV access to more than 4 million acres of state lands conformed to numerous state laws and regulations including vehicle and traffic law, environmental conservation law, DEC rules and regulations, and the Adirondack and Catskill state land master plans. Another goal was to prevent trespass on private lands. Finally, the policy sought to prevent environmental degradation. Commissioner Sheehan emphasized that the policy indicated the commitment of state officials "to managing . . . public land resources in a responsible manner so that they may be enjoyed by current and future generations of New Yorkers." In March, April, and May 2005, the DEC solicited public comment on the new policy, both through public meetings and through letters and email on how best to balance reforestation, conservation, recreation, and preservation goals with ATV use.[28]

The three-month period of public comment generated enormous input. Division personnel have yet to make a final determination but have discerned several basic positions among the many respondents. Conservationists, hikers, and others lamented the excessive impact of ATVs on the environment and desired to limit access. ATVers worried about regulation that may limit their access. Many commentators fell between these two groups. Even among ATVers, DEC employees noticed several groupings. Hunters, fishermen, and older users tend to see the ATV as a tool to provide access to those activities, while younger ATVers tend to see the machine as a recreational device.[29]

Public hearings had a profound impact on the attitudes of many ATVers. At the first hearings they had initially shouted down hikers, made disparaging personal comments, and rudely greeted opposing points of view. This behavior, not the interests of the ATVers, garnered the attention of the media. Several ATVers recognized the need to modify this behavior if the ATV groups were to have the impact they desired on the final policy. They encouraged their members to quiet down, to listen to the opposition, to sit quietly and then respond. "They worked hard at it," said one DEC employee. The ATVers also recognized how well organized other groups were. Hikers and conservationists, for example, had experience over the years in making presentations and dealing with regulators and legislators.[30] In other words, like the responsible snowmobilers who encourage their members to quiet their machines down, responsible ATVers recognized the need to encourage responsible use among their numbers and to work with opponents to reach accommodation.

Creating a new, comprehensive state policy was one thing. Coming up with the financial and personnel resources to follow through on the policy was another. "The first quandary for us," Peter Frank, a DEC employee, told me, was that "we planned ATV use on forest roads and such that were open for automobiles and trucks. But those are public highways and the Department of Motor Vehicles doesn't allow them there. We need separate distinct trails." Frank mentioned that the DEC was "looking for legislation to accompany this policy." The legislation, which failed in the previous legislative session, would have given DEC the power, personnel, and budget to enforce ATV policy. Finally, DEC wants to ensure that people with disabilities can use ATVs as appropriate on public lands.[31]

In such states as Maine with more private than public land, ATV users and their clubs have worked with state agencies to establish and maintain trails. In Maine, ATVs are regulated by the Department of Inland Fisheries and Wildlife as an outdoor sport. ATVs must be registered with the state. The commissioner of the department had repeatedly urged ATV users to show respect for property owners and for others with whom riders share the land. He urged riders to form or join clubs, to encourage others to ride responsibly, to work with owners to gain permission to use land, and to work with law enforcement officials to curb unlawful activities. Unfortunately, some riders were "jeopardizing everyone's opportunity to enjoy Maine's outdoors."[32]

In 1986, in response to growing public discontent with irresponsible ATV use in Maine, the legislature established laws to require that riders secure permission before using any land and that they bear the costs of prosecution for violations of civil or criminal law, and to prohibit ATV use in alpine areas, marshes, and bogs. The law also increased the age at which children could operate ATVs without adult supervision, from ten years to fifteen years, and required training and education programs for those under eighteen. Some Maine agencies seemed unwilling or unable to enforce the ATV laws, in part because ATVs were maneuverable and their operators could easily flee arrest. Nevertheless, the legislation was an important first step toward controlling ATV use—and a much-needed first step at that: according to one survey, three-quarters of Maine residents believed there were major conflicts between ATV users and other interests.[33]

In 1989, Maine officials drafted a new statute to address the facts that law enforcement and educational programs lagged and that environmental damage and trespassing by ATVs were on the increase. This second statute expanded the prohibited areas and gave the Fisheries and Wildlife warden service staff greater resources to enforce the laws. Strangely, legislators reduced from eighteen to sixteen the minimum age for operating an ATV with a required safety program, even though the Consumer Product Safety Commission had publicized the epidemic of pediatric injuries and deaths connected with ATV use.[34]

Ultimately, the revised law proved inadequate, too. Maine ATV registrations grew 136 percent in the ten years ending in 2003 (to nearly 53,000), and retail sales were up 574 percent (to nearly 10,000 annually) in the same period. ATVs outsold snowmobiles. They had a significant economic impact in terms of jobs, fees, and taxes, yet crashes, deaths, injuries, property destruction, illegal use, and other costs had also increased dramatically. Conflicts between snowmobilers and ATVers had turned into war in some communities, as only snowmobilers seemed to police themselves well. The ATVers misused trails, damaging land, polluting streams, and leaving ruts and litter behind. They had only 2,200 miles of dedicated trails versus the snowmobilers' 12,000 miles, which led to epidemics of trespassing. In addition, an ATV public health crisis had emerged. Between 1993 and 2002, 35 people died and 2,241 were injured in ATV crashes in Maine. In 2002, there were a record 319 ATV crashes, with 327 people injured (another record). The six fatalities recorded in that year were the most since 1999, when seven people had died in crashes. On March 18, 2003, Governor John Baldacci convened a task force to deal with the crisis.[35]

Fifteen persons from state agencies and outdoor organizations were chosen for the task force. The governor asked the task force (1) to develop guidelines for a grant program to assist local clubs, municipalities, and landowners in addressing matters of law enforcement, landowner relations, public awareness, safety education, trail development, damage mitigation, and other strategies to solve problems caused by irresponsible ATV operation; (2) to form a subcommittee and work with local, county, and state law enforcement agencies to determine what training, equipment, funding, changes in law, and other resources or actions were needed for those agencies to enforce ATV laws more effectively; and (3) to recommend solutions to ATV problems.[36]

In a series of meetings held in the summer of 2003, the task force solicited commentary from landowners, ATV users, snowmobile operators, law enforcement personnel, and state and municipal officials about how best to deal with the growing controversy over ATV use. Landowners complained that irresponsible users had damaged their property and that they no longer wished to permit ATV users to have access to it, especially since no mitigation fund existed to facilitate repair of damage. They insisted that riders acquire verbal or written permission from landowners for ATV use, not assume it as given. For their part, law enforcement officials worried about being forced to shoulder increasing regulatory, enforcement, and other burdens without the necessary training and financial resources. Many suggested that ATV training be available, but not required, for law enforcement officials.[37]

The task force held four public forums—in Presque Isle, Bangor, Auburn, and Sanford, Maine—and received roughly 180 comments by mail and email. These comments revealed a host of problems facing any effort to regulate ATVs. Citizens wrote about operators tearing down barriers and no-trespassing signs when trails were closed during mud season. Dirt bikers wrote to accuse ATVers of inappropriate behavior (and vice versa). One citizen wrote, "I don't care what law you pass, these are guys who don't care." Another pointed out that prosecution was impossible because violators were "here today and gone tomorrow." Katy Moriarty of the Bangor Water District called for including "public drinking water supply protection areas" among those areas off limits to ATVs, a call that went unheeded. March Perlman, a pediatrician, pointed out that the typical crash cost hospitals between $125,000 and $200,000 to treat. One resident lamented the loss of her brother in an ATV accident and expressed her anger that the operator at fault had been permitted to use an ATV although he had at least two drunk-driving charges against him. Shouldn't ATVs be regulated as if they were automobiles, she complained?[38] How might these concerns be met? According to Paul Jacques, deputy commissioner of the Maine Department of Inland Fisheries and Wildlife, the people who came to the hearings "left their baggage at the door" in search of good policies. As a result, the task force achieved some success, and delivered its report to the governor on December 19, 2003.

The task force concluded that the Maine Department of Inland Fisheries and Wildlife Warden Service should be the lead agency for ATV enforcement, given its existing role as lead agency for boating and snowmobile enforcement. This would require more staff, more equipment, and more training, since there were only ninety game wardens and fifty foresters to handle enforcement along with two thousand state and local police. The wardens required their own ATVs to help in enforcement activities. The task force recommended an extension of grant programs from gas taxes and licenses to provide approximately $120,000 in fiscal years 2004 and 2005. The task force recommended making it a class D crime, with a mandatory $1,000 fine, to attempt to elude an officer, and they endorsed counting ATV violation points and operating under the influence against the operator's automobile license.[39] All of these changes were incorporated into new legislation whose impact is as yet unclear.[40]

The Maine legislation strengthened state programs designed to assist ATV clubs in gaining access to public and private lands, charting trails, and building rights-of-way. The legislature requires trail liability insurance but also subsidizes it, and offers grant moneys to defray the cost of trail development and maintenance. These grants are limited to $2,500. (It says something about the powerful lure of machines that states can provide grants to clubs for building ATV trails but have no money for grants to public interest organizations promoting, say, public health or other social services.) Maine boasts 2,000 miles of trails. While these trails are technically seasonal, for use only from late spring to fall, actual ATV use occurs year round, and operators generally prefer mud season.[41]

Like a number of other states, and with the assistance of the federal government, Vermont helps its citizens to develop trails both for hiking and for motorized use. In 2006 the state estimates it will provide approximately $400,000 to its Recreation Trails Grant Program, administered by the Department of Forests, Parks, and Recreation. Funding comes from the federal recreation trails program and the Vermont recreation trails fund, which are derived from a portion of federal and state gas taxes from off-highway recreation vehicles.[42] Hence, state governments simultaneously encourage ATV use and recognize the need to move them to special trails to limit their environmental impact.

New Hampshire officials have sought a novel way to move ATVs into restricted regions. In November 2005 the New Hampshire Executive Council approved the $2.2 million purchase of a 7,200-acre parcel of land to become a state ATV park with over 350 miles of trails. Located in northern New Hampshire, near Berlin, once a major paper town, the park is intended to bring tourists and their money to an economically depressed area while drawing ATVs away from more congested regions where operators have run into conflict with property owners, hikers, and others. Supporters claim that the park was overdue as a matter of fairness, since "these ATV people have been stiffed."[43]

The Federal Government and Motorized Recreation

Whereas in such eastern states as Maine, New York, Vermont, and New Hampshire, ATV and snowmobile users must negotiate with public and private landowners to develop systems of trails, in most western states, where the federal government is the chief owner of lands, recreationists have little trouble finding a place for internal combustion adventure. Just as they welcomed automobiles and highways, so have the managers of federal land units welcomed ATVs and other recreational machines into the parks. They were tardy both in recognizing that any problem existed with permitting broad access of off-road vehicles to federal lands and in meeting the requirements of federal laws and executive-branch orders. This problem was especially acute in the western part of the nation, where local residents have grown to see the federal government at once as usurping their rights to land and as providing them with subsidized access to grasslands, water, timber, and mineral resources.

Like automobile drivers, ATV operators entered the parks and forests with the strong sense that the use of lands for recreational purposes had the same value and importance as other uses. Operators celebrated their machines as symbols of freedom and the conquest of the frontier. Logically, since they were called "pleasure" craft and "recreational" vehicles, few people considered their use inconsistent with other forms of pleasure and recreation. Given the small numbers of ATVs at first and the vast expanse of land, few individuals initially recognized the potential damage they could do to ecosystems. From the very start, then, ATV use has been considered a right, not a privilege, and the notion has become difficult to dislodge.

A series of laws have long enabled settlers, farmers, businesspeople, miners, and other citizens to gain access to, if not outright ownership of, federal lands. The initiatives concerned grants for the construction of wagon roads, canals, and railroads; homesteading laws; the Mining Law of 1872; the Desert Land Act of 1877; and the Timber and Stone Act of 1878. The General Land Office, established by Congress in 1812 to oversee the disposition of federal lands, was the predecessor the Bureau of Land Management in its efforts to provide access to vast resources to promote the economic well-being of the country—unfortunately, often at subsidized rates and with inadequate oversight that led to a rapacious use of land. In the late nineteenth century, congressmen and senators recognized the need to conserve and preserve some of these lands, creating national forests, parks, refuges, and so on. Yet the overriding interest has always been extraction of such resources as timber, minerals, oil, and gas, which the government has facilitated through road construction.

Road management involves tradeoffs between the benefits of increased access that roads provide versus the associated ecological and economic costs. The National Forest Service manages 10 percent of all public roads in the United States, and its road management decisions will almost always generate controversy because where many people see benefits in access, others worry about the costs. The benefits surely include human safety, firefighting, recreation, and commercial development, while the costs include accelerated destruction of habitat, erosion, impact on wildlife, loss of unique ecosystems, and the high monetary costs of building, maintaining, and decommissioning roads.[44] According to the National Forest Service, 380,000 miles of roads fill the forest system, most of them built over the last fifty years. This period coincides with the rapid growth of recreational machine use. If logging traffic on forest roads tripled over those fifty years, peaking in 1990, recreational use has grown to ten times the 1950 rate. Forest Service personnel acknowledge that funding to manage the system—to maintain it, enforce laws, and meet safety and environmental standards—lags far behind demand.[45]

In considering extraction and harvest to be the crucial component of conservation strategies, many managers have used such short-term economic criteria as jobs created, cubic meters extracted, and products sold as the measure of their success. Many of them failed to study how resource use strategies based on guaranteeing machines access to federal lands have had a negative impact on all other uses, including hiking, and have led to the degradation of most parcels of lands. Other managers of federal land units grasped the essence of the situation only belatedly. Some have been hindered in their efforts to develop workable resource management plans by the allies of recreational machines in Congress, while still others—notably in the case of the administration of George W. Bush—have worked hand in glove with machine recreationists, manufacturers, and their trade organizations to delay reasonable efforts at regulation. Many members of Congress, usually from western states, are loath to force manufacturers to make the machines safer or to restrict their access, and they blame the rider, not the machine, for any environmental disruption. Areas closed to ATVs remain the preponderance, but the areas that ATVs have frequented have been degraded.

One of the challenges in creating limits for ATV use involves the education that land managers received at many public institutions. The ethos of "machine-based progress"—the "machine first" world view of specialists trained to work on federal and state lands—occupies a central place in a doctoral dissertation in watershed management written by John Peine at the University of Arizona. Peine lauded the machine for having enabled Americans to experience nature as it exists far "from permanent human habitation." Americans, he wrote, "are using the products of modern technology to reach into the landscape for a more remote recreational experience with a greater degree of comfort and convenience. The internal combustion engine provides a power source for adventure. The off-road vehicle has come of age." Peine concluded that "the major recreational value of [the operator's] vehicle is in the mechanical development of the machine. The landscape, for him, is a place to evaluate vehicle performance. This type of owner may consider topography to be the most important element of the landscape."[46]

ATV manufacturers fully embrace this view of nature and machine in their advertising. Brochures from Kawasaki, Honda, Polaris, and Yamaha come with the Tread Lightly sticker but clearly do not encourage such behavior. The Kawasaki KFX700 creates "the perfect sand storm." The KFX400 gives "maximum exhilaration." Advertisements for Kawasaki's 2005 sport utility vehicles proclaim that "nothing beats the use of brute force" while depicting a machine perched on rocky cliffs above a beautiful mountain lake in the middle of deep forest. "The hunt for the king of ATVs is over", claims another brochure in which a camouflaged ATV enables a hunter to bag game in a previously inaccessible area of the Rockies. Yet another operator charging through deep mud demonstrates "Big power. Bigger authority." And so on. Again and again, ATV use is depicted as a matter of controlling nature, not respecting it.[47]

Desertification and Dune Buggies

What is the impact of ATVs and other off-road vehicles on various ecosystems? While it took time for the public health costs of such vehicles to be recognized, the environmental impacts were evident from the start. From east to west, in forests and deserts, on public and private lands, whether operating on existing trails or establishing new thoroughfares, ATVs have had an extensive and growing negative impact, usually from the first pass. By the time of Richard Nixon's 1972 executive order, there were five million ORVs (motorcycles, three- and four-wheeled vehicles, and snowmobiles) in the United States. The number of vehicles doubled every two years. They competed for the open spaces, spreading out over rights of way legally and illegally, pushing into forests and plains, creating new trails, moving across streams into a variety of ecosystems.

Scientists alerted citizens and policy makers to the environmental costs of these machines early on, before they numbered in the millions. In 1965 the American Association for the Advancement of Science established a Committee on Arid Lands to work with public, private, and United Nations groups on the problem of desertification. In the 1970s the committee reported on the dangers of ORV use in American (and other) deserts. The entire California desert was at risk because of unregulated ORV use. Plants—some of which took ten years to reach sexual maturity, others that had lived more than one hundred years—were in peril, with the variety and number of species declining as denudation accelerated. Ants and beetles, so crucial to the soil's health, were dying out. Recovery was a doubtful prospect. Erosion was endemic. The aesthetics of arid regions were ruined; tracks and gullies decades old already covered the landscape.[48]

Members of the Geological Society of America convened a study group in the mid-1970s that reached similar conclusions. They wrote that the demand of operators to ride freely "has been largely satisfied by indiscriminate use of Federal lands, and Federal agencies have been generally slow in preventing this invasion of the public domain."[49] The scientists concluded that "damage to wildlife, especially plants, may be irreversible, and many species in the sand dune ecosystem are rare or endangered. The impact of vehicles on dune stability . . . will probably result in increased sand advancement rates into farmlands and lakes . . . freeways and housing developments."[50] Off-road vehicles "have destroyed evolutionary gains of such antiquity that recovery will be exceedingly slow," and "many delicate interdependencies between organisms and their habitats, having been obliterated by ORVs, can never be restored."[51]

Many Americans believe that deserts are devoid of valuable life forms. But over ninety species of shrubs, herbs, and annuals and almost one hundred different vertebrates live in some areas of California's Mojave Desert alone. Vegetation is extensive if not very noticeable at first glance: creosote bush, burro bush, gramma grasses, chickweed, spurge, needle grass, goldenbush, sage, and dozens of other kinds of grasses, bushes, and weeds. Desert shrubs and trees, while commonly sparse, tend to have extensive near-surface root systems. Lichen, fungal, and algal (microfloral) crusts are widespread in arid lands; they are strong enough to protect underlying soil from the impact of raindrops. These organisms have evolved to conserve the limited water supply and to minimize evaporation, to avoid stress in dormant periods and to restrict growth in others, to close leaves at midday, and so on. The soil at least seasonally supports annual plants whose root systems continue to have a stabilizing effect even after the plant has died. Such plants are all quite responsive to human disturbances. It is precisely these soil-stabilizing elements that get ripped apart by ORVs. A single pass by a vehicle can destroy many plants and microfloral crusts; hundreds of passes will destroy them completely, with decades needed to recover.[52]

Desert life forms are important because they contain genetic information on how to survive in harsh environments. Harold Dregne writes that soil "sustains life, functions as a vast reservoir for the collection and storage of water, and absorbs and neutralizes agricultural, domestic and industrial wastes." Yet these desert soils "are probably the most abused" part of the environment. ORVs destroy soil through surface shearing and through compaction of surface soil and subsoil, both of which lead to susceptibility to wind and water erosion, decomposition of what little organic matter there is, weakening of soil aggregate stability, and greater runoff.[53] Compaction of soil extends to 3 feet underground, making the soil hotter during the day and colder at night, crushing air spaces, leading to losses of up to 90 percent of soil moisture along ORV trails, and eliminating vegetation, with the resulting absence of shade accelerating the process. Vehicle use leads inevitably to increased runoff and gullied hillsides.[54] Damage is immediate, extensive, and long lasting if not irreversible, and it contributes to the invasion of foreign species. The geologists urged managers of federal lands to take action immediately, to designate or zone lands to limit use, to register, license, and inspect vehicles, to require ORV operating licenses, to charge use fees to generate income to ensure proper use, and to establish laws and fines including citations, impoundment, and confiscation for trespassing in closed areas.[55]

State and federal officials adopted many of the recommendations concerning fees, licenses, and fines but have been reluctant to limit use—perhaps because they lacked legislative authority or enforcement capability, perhaps because they underestimated how quickly the recreational activity would spread, or perhaps because they had no real sense of the damage. However, there was no reason for federal land managers to remain ignorant of the problem or postpone dealing with it. By the early 1970s the Office of Library Services within the Department of the Interior itself had already turned up an impressive collection of studies that demonstrated conclusively the harmful impact on ecosystems. Cursory review indicated hills denuded, shrubs destroyed, understory annuals wiped out, water quality reduced, and stream bank vegetation eliminated. Any path used for ORV "recreation" (e.g., firebreaks through chaparral) would not revegetate. This meant a loss of resources for grazing, and a loss of habitat for such wildlife as chukar partridge, quail, doves, and cottontail rabbit. Animals suffered not only because of destruction of habitat, starvation, or being hit by machines but also because of noise-related stress.[56]

This evidence notwithstanding, the Bureau of Land Management largely permitted ATVs and ORVs to enter highly fragile ecosystems (e.g., California's Imperial Sand Dunes Recreation Area) with only modest restrictions. The Cahuilla Ranger Station at the Dunes established rules in 2001 to moderate ATV use. They required all vehicles to carry a red or orange safety flag on an 8-foot whip. They limited speeds to 15 mph on public lands within 500 feet of major roads and highways. They restricted the burning of such hazardous materials as gas, oil, plastic, and magnesium. They prohibited glass containers and the use of alcoholic beverages while riding on BLM lands. Given that the machines produce loud noise, the managers outlawed the operation of radios, televisions, musical instruments, and other devices or motorized equipment between the hours of 10 P.M. and 6 A.M. "in a manner that makes unreasonable noise that disturbs other visitors." Public nudity was also prohibited as a nuisance. Though prohibited, nudity and ATVs in an arid environment have reached the public eye in the form of a Playboy video, in which Petra Verkaik plays an archeologist riding an all-terrain vehicle along a dusty road toward a research site. At the site she dances, as apparently some archeologists do, and removes her clothes. Many people who worry about ATV access on BLM and other lands specifically lament the disturbance of Native American burial and other cultural and archeological sites, with or without dancing nude scientists.

Even where lands are extensive and nearby population densities are low, the effects of ORVs have required action. Ray Brubaker, the Wyoming state director of BLM offices, signed off in 1990 on a decision to restrict ORV use in Grass Creek Resource Area in northwestern Wyoming, an area of some 968,000 acres of public land and 1.2 million acres of federal mineral estate filled with unique flora and fauna, located between the Shoshone and Bighorn National Forests and bordering Yellowstone National Park. ORVs had had a substantial impact on the unique geological and paleontological resources of the region, including Native American cultural sites and desert, subalpine, and alpine habitat. (The area would remain open to vehicular traffic associated with grazing and surface mining.) Brubaker closed the Owl Creek, Bobcat Draw Badlands, Sheep Mountain, and Red Butte wilderness areas because of "the very limited semiprimitive nonmotorized recreation opportunities" available in them.[57] By "semiprimitive nonmotorized" recreation, Brubaker apparently meant hiking.

According to the Idaho Sierra Club, ORV use has damaged every type of ecosystem in the nation from the eastern coastal beaches to the mountain ranges, deserts, and beaches of the west.[58] Many people who live near the San Rafael Swell of the Utah Wilderness have tried to get the lands added to the National Wilderness Preservation System. The San Rafael Swell, approximately 33 miles across by 66 miles long and located in Central Utah, marks the northern boundary of Utah's rock desert country, an area that holds some of the largest roadless desert areas in the world.[59] Would protection of these lands put an undue burden on ORV operators? Only 800,000 acres of 9.1 million acres of wilderness-quality land, including Native American cultural sites, are protected as wilderness by law. Already 94 percent of Utah's BLM lands are open to ORVs, while illegal ORV use in protected areas extends the destruction to endangered species, wildlife habitats, and archaeological sites.[60]

Scientists recognize that "wilderness" is crucial to understanding and maintaining complex ecosystems, especially when fewer and fewer such areas exist. A designation of "wilderness" protects native species and their ecosystems. Biologists have shown that, perhaps more than any other technological system, roads accelerate destruction of wilderness and its biodiversity, both by facilitating motorized traffic and by speeding erosion, weed infestation, and degradation of water quality. Human activities and noise on roads disturb a wide range of wildlife. Some animals will change their feeding and other activity patterns to avoid roads. According to conservation biologists, "densities of more than a mile of road per square mile on public land represent a level of human use and access that is incompatible with wide-ranging wildlife such as wolves and bears." They point out that "vehicles venturing off on two-tracks or trails have a well-documented role in creating seedbeds for weeds and promoting their dispersal." In the intermountain West, roads facilitated the spread of exotic weeds that outcompete native plants, with adverse effects up the food chain in pronghorn antelope, deer, small vertebrates, native birds, and insects. For these reasons, specialists favor prohibition of ORVs in public lands to protect the health of native species and natural communities.[61]

The accuracy of Bob Marshall's predictions about what happens to wilderness and parkland when roads are built and machines arrive has become starkly clear through the Starkey project. The Starkey project is an ongoing long-term project to study the impact of resource use in national forests on mule deer *(Odocoileus hemionus)* and elk *(Cervus elaphus)* populations and habitats. The Starkey Experimental Forest and Range includes 40 square miles on the Wallowa Whitman National Forest in Oregon. Much of this research represents the first attempt to measure such pressures on elk and deer as hunting, ATV and other human disturbances, human densities, traffic rates, and other variables. Seventy years after Marshall ruminated about the impact of roads on wilderness, managers and biologists believed they still lacked sufficient data to ruminate about ungulates. The growing demand to use forest resources for multiple uses—intensive logging for home and other construction and such recreational activities as hunting, hiking, horseback riding, and machine recreation—called for long-term scientific study. Entering the twenty-first century, managers lacked sufficient longitudinal data to evaluate the situation objectively. How were they to struggle with the political pressure to use potential wilderness areas for grazing, agricultural, and extractive businesses and as playgrounds for 500-pound 200-hp machines? The Starkey project results, which indicate that the legacy of national parks and forests is being threatened, provide precisely these data.

The Starkey researchers planned and carried out the project carefully. The planning stage took four years; it took another four years to establish the research facility, then five to seven years to complete the initial studies. The research facility required the construction of one of the largest ungulate-proof enclosures ever constructed. It became operational in 1989 and had a "novel, automated radio-telemetry system that could track the movements of more than 100 radio-collared ungulates" twenty-four hours a day. The Starkey project involved "all groups with strong interests and investments in management of national forests," including state and federal wildlife and land management agencies, timber companies, livestock associations, tribal nations, and conservation groups. Universities joined in to take advantage of research opportunities for graduate students. Over one hundred scientists have been involved in the studies,[62] the first of which are now being published.

One study, for example, found that roads and the hunting they facilitate have hurt such species as elk and mule deer, both targeted and nontargeted animals, whether the animals hid or ran from hunters. Those that successfully eluded hunters by running may have depleted fat reserves and suffered energetic costs that put them at risk, especially in winter. These animals moved abnormally long distances and otherwise changed their behavior.[63] Intensive timber harvest has also had a significant impact on elk and cattle. The intensive timber management practices of the latter half of the twentieth century changed habitat, in some cases irreversibly.

Even short-term disturbances by timber harvest, concomitant road building, and resulting traffic caused elk and other ungulates to change their behavior. Some migrated as far as 5 miles. While some studies showed that timber harvesting produces forage areas, continued harvesting is required to maintain these new forage areas. The alternative was to permit regeneration of coniferous forests. The average number of days it took for hunters to "harvest" an animal declined during and after timber harvest. Hunters found it easier to hunt owing to roads and machines.[64]

In response to the 1972 executive order, and to the growing use of ORVs on national lands, the Bureau of Land Management, the Forest Service, the Fisheries and Wildlife Service, and the National Park Service were required to undertake environmental impact studies about whether and how to limit ORV use. They presented the results to the public for comment. Then they determined how much of the land to close to ORVs and where to allow them to be ridden. For example, in the Wenatchee National Forest, a parcel of 2.2 million acres stretching 135 miles north to south along the crest of the majestic Cascade Mountains, with dense forests, lakes, rivers, and streams, forestry specialists set out to regulate ORV use in the 1970s. As of 2004, 40 percent of the forest is designated as wilderness, with no vehicular traffic allowed.[65]

At the time of their first environmental impact study in 1976, the Wenatchee rangers permitted use of ORVs in virtually all parts of the forest unless a prohibition was posted. Study of the impact on soils, water quality, wildlife, and recreation led them to consider five alternative regulations, each one more restrictive than the last. The rangers did not offer an opinion on which alternative they preferred, to ensure public involvement in the resolution.[66] It's hard to fathom their hesitation, given the fact that their study had showed that ORVs "disturb and loosen the soil surface, making it very susceptible to both wind and water erosion." Rangers observed that "vegetative disturbance may be caused by off-road and off-trail use by motorized vehicles." ORVs may diminish "water quality by increasing sediment loads through soil disturbance and by the addition of pollutants into streams or water bodies." They acknowledged that noise might bother or disturb nesting or calving wildlife, a "temporary" problem that could be solved by muffling. They also noted that "there will be a temporary impact on air quality," including dust, emissions, and odor. Despite this litany of ill effects, the rangers somehow concluded that "no permanent long range impacts would be expected [to result from ORV use]. Sufficient vegetation and air movement is available to offset carbon-monoxide–oxygen exchange and neutralize exhaust emissions."[67] Where, one wonders, did they find evidence that the Wenatchee forest was self-cleaning? At Wenatchee and at other federal land units, each manager seems to have chosen slightly different criteria to evaluate the impact of ORVs and to determine whether to restrict them, and in many cases the required evaluation was never completed at all.

Congressional involvement in these considerations took on greater significance after the completion of the Public Land Law Review Commission study in 1976. In the wake of the publication of Rachel Carson's *Silent Spring* (1962), Congress passed several crucial laws and acts to protect the nation's land and other resources: the National Environmental Policy Act (1969), the Endangered Species Act, the Clean Water Act, the Clean Air Act, the Resource Conservation and Recovery Act, the Wilderness Act, the Wild and Scenic Rivers Act, the National Historic Preservation Act, and the Federal Land Policy Management Act, or FLPMA (1976). The FLPMA stipulated that "public lands be retained in Federal ownership, unless as a result of the land use planning procedure provided for in this Act, it is determined that disposal of a particular parcel will serve the national interest." The act also stipulated that BLM manage public lands for "multiple use."

The commission, which served from 1964 to 1970, examined well-established ORV use and animal grazing practices. It undertook a comprehensive review of public land laws and the rules, regulations, policies, and practices of federal, state, and local governments. The commission grew out of a letter dated October 15, 1962, from Wayne N. Aspinall (D-CO), chairman of the House Interior and Insular Affairs Committee, to President John F. Kennedy suggesting the need for a broader examination of public land policy and inviting submission of the president's views to the Eighty-eighth Congress. The president concurred and instructed the secretaries of the interior (Stewart Udall) and agriculture (Orville Freeman) to represent him. On August 14, 1963, Aspinall introduced a bill to establish the Public Land Law Review Commission. On September 19, 1964, President Lyndon Johnson signed Public Law 88-606 establishing the nineteen-member commission. Aspinall was named chairman; six members were appointed by the president, six by the U.S. Senate, and six by the U.S. House. The Senate and House members changed from time to time, because of election defeats and resignations. Congressman Morris K. Udall served on the commission from 1967 until 1970, when the final report was issued and the commission disbanded.

The report addressed a variety of land issues: acquisitions and exchanges; administrative procedures; Alaska; disposal of public lands; economic impacts; energy and nonfuel minerals; timber; water; user fees and charges; the environment; fish and wildlife; grazing; agriculture; land grants; recreation. The commission members recommended "retaining [land] in Federal ownership whose values must be preserved so that they may be used and enjoyed by all Americans," a position formally adopted in the FLPMA.[68] This position triggered the so-called Sagebrush Rebellion in which several counties in western states enacted ordinances prohibiting BLM land managers from entering or taking action on federal lands, based on the assertion that the states owned the lands.

The Sagebrush Rebellion gained momentum in the 1960s, especially in Nevada and Utah. In Nevada, 87 percent of all land is federally managed and controlled. In opposition to the Public Land Commission recommendations, Nevada legislators created a Select Committee on Public Lands to seek changes in public lands policies. They sought the cooperation of western state and local governments, an effort joined by the Western Council of State Governments and the Western Interstate Region of the National Association of Counties, leading to the formation of the Western Coalition on Public Lands. Members of the coalition believed that federal policies affecting the west were made in ignorance of local conditions and concerns, that policies were made for a "national" constituency without regard for western problems, and that this disregard would intensify in the effort to satisfy the nation's energy needs, in the control of access to grazing and mining development, in military land withdrawal, and in the closure of selected public lands to hunting and fishing.[69] Many Sagebrushers were ORV enthusiasts.

The revolt of state legislators triggered federal legal action that overturned those ordinances. But efforts to weaken BLM authority vis-à-vis the states continue from within Congress itself. Western senators and representatives have sought to weaken BLM power by refusing to reauthorize programs or budget for them. As a result, BLM managers have faced difficulty in recording valid mining claims, issuing permits, granting rights of way, restoring riparian zones, developing or amending land plans, fighting fires, selling and disposing of lands, issuing or renewing grazing permits, preparing timber sales, conducting surveys, controlling noxious weeds, undertaking environmental activities, and so on. They have faced growing pressure from ORV operators who wish unimpeded access as one of those "multiple uses."[70]

The efforts of local officials to take back land and make the roads in national monuments and wilderness areas serve machine-driving westerners have not abated. In Kane County, Utah, where President Bill Clinton established the 1.7-million-acre Grand Staircase–Escalante National Monument, local commissioners claimed ownership of hundreds of miles of dirt roads, dry washes, and riverbeds. Wrote one investigative reporter, they "graded roads and put up fiberglass signposts, inviting all-terrain vehicles onto federal lands. The explosion in the use of those vehicles, whether motorized dirt bikes or four-wheelers, has left its echoes and tracks on a landscape where only people, horses and burros went before. Such trails strengthen local claims to roads, whose very existence may be disputed by others." In Kane County, federal signs stand alongside county signs, the former prohibiting the use of ATVs, the latter inviting them in. The commissioners sought to encourage wilderness use by tourists. The destruction of federal property is a crime, yet county commissioner Mark Habbeshaw, allegedly the culprit behind the signage and the violation of the law, faces no charges. In Washington, Richard Durbin (D-IL) announced his intention to hold up the nomination for the number two post in the Interior Department over the Kane County dispute.[71]

One reason for the difficulty in reaching consensus about ATV use on federal lands has been the increasing politicization of science in the past fifteen years. Scientific activity should not be a highly paid lobbying effort to direct legislative action in support of industry or against it; rather, it should concern the testing of hypotheses with evidence. In the postwar years a technology assessment process evolved that required promoters of such technologies as dams, reactors, industrial processes, drugs, and so on to demonstrate that they could be used safely. Before the 1960s, opponents of a particular technology had the onerous burden of proving it unsafe. Several challenges impeded their gathering and evaluation of evidence. First, manufacturers rightly cited proprietary rules as a reason for not sharing data. Second, the negative consequences of a technology often emerge only after some period of time. Third, research protocols present their own sets of uncertainties (e.g., it is difficult to extrapolate from animal studies to human impacts). These challenges frequently put an unfair burden on opponents. Yet both opponents and promoters largely found the new technology assessment process that evolved in the 1970s, with its requirement that promoters produce environmental impact statements and other studies in support of their position, to be workable and fair.

Unfortunately, in too many cases officials in federal land management units have failed to conduct studies, have ignored evidence, or have used administrative rule- and standard-making procedures to avoid making the hard but perhaps reasonable decision to restrict ORV access to federal lands. This last path avoided disappointing the increasingly vocal community of recreational machine users, but it disappointed hikers, hunters, environmentalists, and many scientists. Other officials, connected too closely to industry, have sided with manufacturers in their costly demands to conduct study after study in search of at least some evidence demonstrating the possibility of consonance between machine and nature. In the meantime, more machines enter the desert, forests, and grasslands.

Federal Inattention to ATVs

In what should have been a devastating report released some twenty-eight years after the first executive order, the General Accounting Office (GAO, the investigative arm of the U.S. Congress) released a study in 2000 that documented the repeated failure of managers of federal lands to limit in any substantive way the use of off-road vehicles, personal watercraft, and other recreational machines. The study, based on a survey of twelve hundred personnel in the four federal land management agencies that were responsible for 95 percent of all federal lands, indicated largely unregulated use by fourteen million owners of the off-road machines in 1999 despite evidence that such use damaged land, plants, wildlife, and other resources and generated significant conflict with other users.[72]

Managers in the Bureau of Land Management, the Fish and Wildlife Service, the National Park Service (of the Department of the Interior), and the Forest Service (of the Department of Agriculture) showed disturbingly varied awareness of the dangers that recreational machines posed. GAO researchers asked them, Do the lands and waters in their units have the capacity for personal watercraft (PWC) or snowmobile use? They defined "capacity" for the former as having "any water on or adjacent to the lands administered by the federal unit that support or could potentially support their use," and for the latter as having "suitable terrain and sufficient snow depth in an average year to operate these vehicles." Managers reported that in 1999, PWCs and snowmobiles were used in 475 of the 1,018 federal units (47%) that responded, with a range of 31 percent in national parks to 82 percent in forests. While such users may have accounted for a relatively small number of total visits, they are the heaviest users of resources, and in some periods they constitute a significant proportion of total visits (e.g., 43% of winter users of Yellowstone National Park are snowmobilers).[73]

The GAO investigators concluded that in making determinations about which restrictions, if any, to apply, the managers of the units "often do not have any information on the impacts" of these machines on resources and environment. While a variety of laws and executive orders authorized them to monitor their impact on resources, safety, and other users, 60 percent of them had not collected enough information to do so, and of the remaining 40 percent, about half said the information was inadequate for determining how to manage use. Given the failure of the federal government to fund the national parks and recreational areas at a level permitting repairs, maintenance, and upkeep, let alone law enforcement, it is no wonder that monitoring and data collection got short shrift. Accordingly, the authors of the GAO report recommended that the secretaries of the interior and agriculture have their units evaluate and monitor impact and use that information in making future decisions about whether to continue to allow this use and, if so, how that use should be managed.[74]

Park Service and Fish and Wildlife managers generally prohibited the use of recreational machines unless it was demonstrated that no harm was likely to result to the resources or environment. By contract, Forest Service and Bureau of Land Management managers generally allowed their use unless the unit manager clearly demonstrated potential harm. Generally, the machines were prohibited in wilderness areas and specifically authorized in other areas such as national recreation areas. If no law either prohibited or authorized such use, the federal agency responsible for managing the area made a determination on a unit-by-unit basis.[75] This seemingly democratic approach led to ad hoc, unscientific determinations of usage patterns. In many cases the policy was no policy at all. Rather, operators used parks and forests as they wished, and managers remained silent.

In 2003, for example, in the Chattahoochee-Oconee National Forest in Georgia, the Forest Service dropped a proposal to open 100 miles of roads for ATV use. ATV users already had access to 133 miles of trails in the Oconee and Chattahoochee forests, and in the entire national forest system, covering more than 190 million acres in 155 forests, only two forests prohibited ATVs. Worse still, irresponsible and illegal use far exceeded responsible and legal use. Managers in Chattahoochee discovered in fact over 550 miles of illegal trails, some in designated wilderness areas and on paths restricted to pedestrians. The estimated cost of closing, rehabilitating, and revegetating those trails was $1 million. The Forest Service was at a loss how to deal with the menace. Foresters in Wayne National Forest in Indiana repeated the litany of illegal ATV use: "Whether we look at the designated trail system or the non-ORV management areas, we have no control over off road vehicle use. We install signs and they are ripped out. We erect barriers and they are removed or ridden around. We rehab areas and they are violated again and again."[76]

The inattention of officials in the Forest Service and other federal bureaucracies to illegal encroachment by ATVs has generated concern among various groups—public and private, hunting, conservation, and other—that have resorted to legal action to require the officials to protect the nation's remaining wilderness areas.[77] The Texas chapter of the American Fisheries Society advocated restrictions on ATVs because they destroyed publicly owned streams and river habitats, threatening biodiversity, eliminating the more "desirable" species such as sport fishes, and ruining fishing opportunities.[78] In the meantime, various trade organizations and clubs vigorously defended the rights of ATVers, dune buggy operators, and other enthusiasts to "the use of public lands." They attacked the alleged "junk science" of groups seeking to preserve wilderness. For example, the president of the American Sand Association denounced attempts to limit access to millions of acres of desert, his organization having secured motorized access to "only" 2 million of 25 million acres in the California Desert Conservation Area (or 3,125 square miles, twice the size of Rhode Island).[79]

Hikers, hunters, and other outdoor enthusiasts reject the argument that efforts to protect the environment are "frivolous" or based on "junk science." The noise and pollution shock them, the soil worn away from the hillside angers them, the irresponsible use stupefies them.[80] Sportsmen (and sportswomen) generally welcome technological advances that facilitate access to the out-of-doors. Most of them also worry that regulation of technology may be the first step down the slippery slope of encroachment into what they believe—and the courts have recognized—is the right to bear arms. But many recreationists have come to see ATVs as nuisances that shake the forest floor, frighten the animals, leave a trail of fumes, and permit individuals who are not real hunters to pursue game on machine-back, often running those animals into the ground.

In the face of lagging federal activity to ensure safer, cleaner ATVs, state governments have acted. In California, such vehicles are typically used adjacent to or in urban areas that already suffer from poor air quality, for example in Hungry Valley (with impact on Los Angeles), at Pismo Beach (adjacent to San Luis Obispo), and at Ocotillo Wells (which contributes to San Diego's air pollution). Working closely with industry, California officials developed reasonable emission control regulations for ATVs, jet skis, lawn mowers, chain saws, golf carts, and the like. The regulations required that these machines be equipped with catalytic converters, fuel injection, and other technologies, but allowed old equipment to continue being used and replacement parts to remain available for them.[81] In 2003, California regulators followed up with new noise standards for ORVs operated in state vehicular recreation areas that reduced noise emissions from 101 dB to 96 dB (96 dB is hardly quiet, though, nor is it the best that leading manufacturers can do).[82] The success of industry and regulators in California indicates that, without waiting for additional study and without affecting their ability to sell ATVs, manufacturers can build machines that are more environmentally sound. Will the impetus to deal with a vast and growing public health crisis connected with ATV operation come from medical personnel, consumer groups, and government officials, or from manufacturers?

Public Health and ATVs

ATVs have several qualities that make them so attractive. They are fast and maneuverable. They bring joy to club members. They enable people with a love of the outdoors to gather and ride together. ATVs give new meaning to wilderness. They take operators to vistas they otherwise might miss. For all the same reasons, ATVs are also extremely dangerous. They encourage millions of people, experienced and inexperienced alike, to ride unstable machines on difficult terrain. ATVs have killed thousands of Americans and injured hundreds of thousands of others, a disproportionate number of them operators sixteen years old and younger. Having failed to establish meaningful restrictions on use of ATVs on federal lands, government officials belatedly turned to the issue of ATV safety. Here too it seems that a mistaken sense of individual rights and reluctance to regulate business activities has triumphed over common sense, safety, and justice.

Just as in touch football, softball, and other sports, the weekend warriors on ATVs are the ones most likely to wake up with bumps and bruises on Monday morning. Often they wake with broken bones, contusions, and lacerations. And all too frequently they don't wake up at all. Alcohol, overconfidence, and excessive speed usually contribute to the accidents. On the way back from a visit to the Polaris factory in Roseau, Minnesota, I shared restaurant space with two men who typified this phenomenon. Upon hearing of my interest in recreational machines, both offered stories of their escapades. To my left sat a twenty-eight-year-old man who considered ATV riding a wonderful experience but second to the exhilaration of snowmobiles. He preferred moving at 70–80 mph on the long, flat, smooth stretches of frozen rivers and lakes. Snowmobiles were safe, he said, and quite stable, although he had flipped two of them, fortunately without ill effect to himself or machine. The thirty-two-year-old to my right had not been as lucky and was therefore more circumspect. On his first ATV ride as a thirteen-year-old, he had flipped a three-wheeler, breaking bones in both arms and losing skin on one side of his face. For a time during the healing he had been unable to see out of one eye, because of swelling and because the oozing scabs had glued the eyelid shut.

By 1984 the Centers for Disease Control (CDC) and the Consumer Product Safety Commission (CPSC) recognized that the number of ATV accidents had skyrocketed. ATV use had become increasingly deadly, an increase not fully explained either by the growing number of ATVs in use or by better reporting to the CPSC. Children under the age of sixteen accounted for 37 percent of the total estimated injuries from 1985 through 2001; children under twelve years of age accounted for 18 percent of the total injuries. The CPSC urged caution for three- and four-wheeled ATVs as the number of injuries climbed tenfold over five years, from 8,600 in 1982 to 86,400 in 1986, with a total of 696 ATV-related deaths in 1982–87. Three hundred thirteen of those deaths (45%) involved children under sixteen years old and 20 percent under twelve years old. The CPSC recommended that children under age twelve should not operate ATVs at all, that children between twelve and fifteen should not operate an adult-size ATV (greater than 90 cc), and that hands-on training courses should be required of all operators. A total of 2,414 deaths associated with three- and four-wheel all-terrain vehicles occurred from 1982 to 1993. In 1990 there were 2.75 million ATVs in use, with half being three-wheeled machines sold before 1986. An in-depth study of death statistics by the CPSC estimated that perhaps 90 percent of accidents were not survivable even with immediate emergency care.[83]

A high frequency of ATV accidents was clear from the first. In Alaska alone, over a two-year period from January 1983 through December 1984, accidents involving three-wheeled ATVs caused 20 deaths and 538 injuries, including six persons permanently disabled by neurological injuries. Fifty-five percent of the fatalities were among males; 75 percent of the deaths occurred among people aged fifteen to thirty-five. Ten persons died as a result of head trauma. Only two operators wore protective helmets, but helmet use merely lessened the risk of death or serious injury; it did not eliminate it. Officials estimated that the cost of inpatient care for ATV accidents in Alaska was $1.6 million (in 1983–84 dollars), and the cost to care for the permanently disabled individuals would be $11.5 million were they to live to age sixty-five.[84] Given the roughly three hundred to five hundred Americans killed (data are incomplete) and the more than one hundred thousand hospitalized annually from ATV use, and with rising health care costs far outstripping inflation, we must ask if these are reasonable costs and how citizens should expect to pay for them.

Canadian physicians took the lead in studying the growing plague of pediatric orthopedic injuries associated with ORV use. They studied ninety boys and forty-three girls who had musculoskeletal injuries related to ORV use and were admitted to hospitals in the two largest urban centers in Manitoba between April 1979 and August 1986. Dirt bikes were implicated in ninety-three admissions, snowmobiles in seventy-two, and ATVs in sixty-eight. The average duration of stay (in days) was twice as long for snowmobiles over dirt bikes and 60 percent longer for ATVs. There were 352 fractures of an extremity or the spine.[85]

From 1985 through 1997, the CPSC identified 113 deaths associated with ATVs in West Virginia, with approximately two-thirds of the deaths caused by injury to the head or neck. Consistent use of helmets by riders might have reduced ATV-related deaths substantially.[86] Another study revealed that ATV-related facial trauma had increased significantly, especially in pediatric patients, who accounted for 39 percent of all ATV fatalities. Physicians at a Pittsburgh hospital had treated thirty-five referrals from western Pennsylvania, Ohio, and West Virginia between 1988 and 1991, all of whom had suffered injuries from ATV use. Virtually all accidents were off-road. Eighty percent of the patients were male. The patients ranged in age from twenty-three months to eighteen years, but predominantly from eleven to eighteen. In 57 percent of the cases the injured party was the driver. The most common kind of accident was flipping or rolling over (63%) followed by a collision (20%). The physicians noted not only the severity of the accidents and their frequency but also the fact that ATV use "necessitates a high level of driver interaction with the vehicle, often requiring adjustment of weight distribution," which children apparently found challenging.[87]

Most state officials have been hesitant to require helmet use, however, deeming it a paternalistic infringement on personal liberty. Only twenty-three states have helmet laws, but many of the laws require helmets only for ATVs used on public lands or for children, and many states repealed their helmet laws even as ATV deaths skyrocketed. Between 1982 and 2002 there were over 4,500 ATV-related deaths and 1.4 million injuries that required hospital treatment. These data were likely incomplete, for several reasons. First, the CPSC and the National Transportation Safety Board did not systematically receive or collect data on ATV injuries and mortalities until 1999. Second, each state had different, and occasionally inefficient, methods of collecting accident and mortality statistics. Third, after the introduction of a new vehicle, emergency room personnel often did not recognize and therefore report accidents or mortalities as attributable to a specific technology. But these data problems did not obscure the fact that ATVs were dangerous to life and limb.

When CPSC statisticians began to employ a new National Electronic Injury Surveillance System with the goal of producing accurate data, they recognized that the total number of deaths and death rates per ten thousand ATVs were significantly higher than reported. The system indicated that in 2000, 547 deaths had occurred versus 344 reported, and the risk of death per ten thousand four-wheel ATVs in use had risen to 1.5. Indeed, after declining steadily from 1985 until 1993, a period that coincided with the phasing out of three-wheeled ATVs, the number of injuries associated with ATVs climbed (see table A.4).[88] Specialists estimated the cost of ATV-related injuries among children seen in emergency rooms from 1992 to 1994 at $643 million (for 93,207 injuries, with an average cost per injury of $6,899). In 1994, four-wheeled ATVs accounted for 67 percent of all ATV-related fatalities. Again, children suffered heavily: 942 of those killed (37%) were under the age of sixteen and 406 (16%) were under the age of twelve. Fifty percent of the deaths took place in twelve states. The five states with the most fatalities associated with ATVs have been California, Pennsylvania, New York, Michigan, and Texas.

At the end of the 1990s researchers observed another significant spike in ATV injuries and deaths. The CPSC reported a statistically significant increase in the number of injuries (20%) in 1999 over 1998, with 35 percent of those injured being under age sixteen and 15 percent under age twelve.[89] The CPSC analysts concluded that as many as 20 percent of accidents had gone unreported in recent years. There had been hundreds of thousands of hospitalizations, between fifty thousand and eight-five thousand annually, with the hospitalizations increasing along with the numbers of both users and machines, and the risk per ten thousand ATVs was increasing as well.[90] The CPSC annual report of May 2002 showed "a significant increase in the estimated number of injuries for 2001, up about 17% from 2000," that could not be explained by increasing numbers of ATVs in use, and a vast increase in the number of deaths since the previous report, perhaps due to the fact that since 1999 more complete data on public road fatalities had become available.[91]

Fighting the demands of "anti-ATV forces" for a "dumbing down" of ATVs with roll bars, seatbelts, and other "so-called safety measures," the Specialty Vehicle Institute reported that there had been "a 5 percent decline in the ATV injury rate from 2001 to 2002; a 31 percent decline in the injury rate from 1988 to 2002"; a sharp drop in the fatality rate from 1999 through 2001; and a sizable decline in the proportion of total ATV-related injuries that involved children under age sixteen since 1997. Institute officials drew these numbers from analysis of a CPSC report to claim that "the numbers aren't as bad as ATV opponents suggest." Sales of ATVs had increased steadily in the United States since 1991, but the CPSC data showed that "the number of injuries is growing at a lesser rate in proportion to the ATV population itself." Tim Buche, president of SPVI, said, "The decline in injury and fatality rates show that the industry's focus on ATV safety programs is working, but that more cooperation is needed to help pass appropriate state ATV safety legislation to further reduce ATV-related injuries."[92] The CPSC report had prompted claims by activists and consumer groups that Buche characterized as inaccurate and misleading. "Raw numbers don't tell the whole story," he said, "especially when key data is omitted because it doesn't fit a particular organization's agenda. The safety of our customers is of paramount importance to the ATV industry, and as far as we're concerned, even one injury is one injury too many." He continued, "Rather than attempt to mislead the public with inflammatory and inaccurate claims, we urge consumer groups to focus on promoting rider safety through appropriate state ATV safety legislation, rider education programs, and parental supervision."[93] Nevertheless, even if accident and death rates have declined as a percentage of the (growing) number of ATVs, the absolute numbers of deaths and hospitalizations remain high. The increase in casualties has slowed, but it has not stopped.[94]

Is rider education the answer? Consider the sixty-three-page *Tips and Practice Guide for the ATV Rider* published by Honda and the ATV Safety Institute. "Stupid hurts," the brochure warns, and a note on the inside front cover acknowledges that ATVs "may present a risk of death or severe injury in certain circumstances." But the language of the safety instruction that follows is strangely languid, as in this description of how to avoid being crushed on a steep hill: "When going uphill keep your weight uphill. Never allow the ATV to roll backward. Dismount uphill or to a side if pointed straight up hill. If you roll backward, keep weight uphill and apply front brake. When you come to a complete stop, apply the rear brake. If the ATV continues to roll backward, dismount uphill side immediately." Anyone who has survived an ATV accident knows that once the machine begins to move in an unintended direction—once it begins to roll, once it tips—there is little that even the most experienced operator can do.[95] Owners' manuals, fine print, and voluntary processes cannot do the job of protecting ATV users. According to consumer and other analysts, ATVs are inherently unstable—they have a high center of gravity—and are risky to operate at the speeds for which they have been built and the terrain for which they are intended.

Fits and Starts of Federal Regulation

During the 1980s, as deaths and injuries mounted, officials of the Consumer Product Safety Commission sought to regulate the sale and use of ATVs to protect their operators, especially underage operators. The manufacturers of ATVs, for their part, were determined to avoid controls. To avoid litigation, the CPSC eventually negotiated a voluntary consent decree from the manufacturers to cease sales of three-wheeled ATVs; to cease sales of ATVs with engine size greater than 90 cc for use by individuals under the age of sixteen years; to establish a refund program for the return of adult-sized three-wheeled ATVs used by children; and to commence informational safety programs. The decree did not actually remove three-wheeled ATVs from the market; it still permitted their resale. The parties agreed to an industry-sponsored $100 million safety program offered free of charge to distributors and purchasers, a program that probably contributed to the 33 percent decline in injuries between 1984 and 1988.[96]

Nevertheless, as in many other cases involving the regulation of a dangerous technology or product (e.g., cigarettes, asbestos, pharmaceuticals, pesticides), this negotiated consent decree principally allowed manufacturers to avoid fines, escape liability, and put off a resolution of safety issues. On September 29, 1987, members of the Senate Committee on Government Operations produced a report that criticized the CPSC and the Department of Justice for failing to act more decisively to protect U.S. citizens from the dangers of ATVs. The report found that "use of ATV's presents both an unreasonable and an imminent risk of death and serious injury requiring immediate enforcement action by the CPSC."[97]

Several of the committee members believed that the CPSC had simply let the ATV industry off the hook at a time when deaths were running at twenty each month. The committee also blamed the Department of Justice for its "baffling and unconscionable delay" in taking action and for its failure to act even in the face of fourteen thousand pages of evidence and a $2 million investigation by the CPSC. In a majority report, the Senate committee urged the CPSC to initiate legal proceedings to control an unsafe product in any future case involving more than ten days of Justice Department delay. The committee members also recommended that the Consumer Product Safety Act be amended to require that suit be brought within thirty days after any CPSC determination of an "imminent hazard" to public health.[98]

Although the evidence indicated that ATVs were killing Americans and putting others in the hospital, such Republican members of the committee as Jim Lightfoot, Larry Craig, Buz Lukens, then–Speaker of the House Dennis Hastert, and Senator James Inhofe defended the ATV manufacturers. They professed sympathy for those who had lost loved ones but worried that a decision might be made on the basis of incomplete data. Since half of the ATV-related deaths had resulted from collisions, they reasoned that driver error, not machine design, was the source of the problem. They also argued that there was no evidence to support the claim that a recall would reduce the number of deaths and injuries. They supported the Justice Department's delay in taking action against manufacturers because the CPSC had lost suits over recalls in the past because of evidentiary flaws—at great cost to taxpayers. They concluded that the ATV was not "an imminent hazard."[99] Apparently hundreds of thousands of hospitalizations and hundreds of deaths, including deaths of children, were simply to be expected as a part of recreation in the United States.

Three years later, under the leadership of Senator Joe Lieberman (D-CT), the Government Affairs Committee again took up the matter of ATV safety. The weak consent decree negotiated between the government and the manufacturers of ATVs had only briefly slowed the pace of injury and death. Unscrupulous dealers continued to sell the adult machines—those with the more powerful engines—to underage users. For example, the Massachusetts Public Interest Research Group selected ATV dealers at random from the Yellow Pages and found that over half of them would sell adult machines for use by children. The committee determined that, by their very nature, one in three ATVs would be involved in a serious injury or fatality during the vehicle's lifetime. Already, over half a million Americans had been injured by them. Lieberman therefore called for making it illegal to sell adult-sized ATVs for use by children; instituting a permanent ban on the sale of three-wheeled ATVs (at that point the ban extended only until 1998); and establishing a final recall and refund program for the million-plus three-wheeled ATVs still in use (Lieberman pointed out that the federal government had insisted on a recall of the Ford Pinto in the 1970s after its gas tank design had led to sixty-one deaths). Lieberman also urged the CPSC to establish safety standards to prevent lateral rollovers. He concluded, "If, as the ATV industry contends, a meaningful lateral stability standard is not feasible and would not reduce injuries, then CPSC must determine whether ATVs can, in fact, be made safer and, if not, whether a partial or complete ban is warranted."[100]

A series of witnesses supported Lieberman's conclusions. Charles Chvala, a state senator from Wisconsin, testified that "federal action is needed now more than ever. ATVs are kid-killers, plain and simple." Every year a dozen or so children died in ATV accidents in his state.[101] Dr. Mark Widome, the chairman of the committee on injury and poison prevention of the American Academy of Pediatrics, said that "ATVs pose a particular problem because they are inherently unstable. They present to the average user an unacceptably high risk of personal injury. The vehicles are prone to roll over and to crush the rider. There is perhaps no other popular recreational activity which is so demanding, demanding so far in excess of the average child's capabilities, and yet is so unforgiving of any error or misjudgment."[102]

Mary Ellen R. Fise, the product safety director of the Consumer Federation of America, criticized the foot dragging in the government and among the five manufacturers. They had reached an agreement to establish voluntary industrywide safety standards within four months, yet that period had stretched to two years, at which point the industry reported that it was unable to develop a lateral stability standard. Fise observed, "Meanwhile, four-wheel ATVs continue to flip over and kill or injure their riders." The manufacturers had introduced such new children's products as an ATV that weighed 97 pounds with a top speed of 34 mph. With incredulity, Fise commented, "Clearly, what we don't need are more vehicle sales that will allow our 7-year-old children to travel 34 miles per hour and risk being smothered or pinned down by a 97-pound machine."[103]

Robert R. Wright, former dean of engineering at Ohio State

University, the president of his own engineering firm, with extensive experience in litigation involving accident reconstruction, product liability, and human factors considerations, also provided testimony. Wright had studied over two hundred accidents involving both three- and four-wheeled ATVs. His findings were an indictment of ATV safety based on simple physical laws: a vehicle with a high center of gravity is less stable than one with a low center of gravity. Wright testified that ATVs "are defectively designed and are very susceptible to accidents involving sideward, backward and forward overturns. The current design of the present three-wheel, and in many cases the four-wheel machine, has the rider sitting high, so the center of gravity of the system is high in relation to its track width and its wheel base. This design defect is the cause of a large percentage of ATV accidents where individuals were severely injured or killed." Operators managed this inherent instability by shifting their weight, but in an emergency, Wright observed, there would be no time to perform the necessary physical gyrations. Industry tests indicating that the machines were stable—tests on the basis of which manufacturers wanted to promulgate a lateral stability standard—used expert operators, not average ones. Wright argued that if ATVs were tested with a 200-pound weight attached to the seat, similar to the ANSI standard for riding lawn mowers, the results would indicate that "the present ATV standards are sub-par and inadequate."[104]

Industry representatives—eager to continue selling these increasingly popular machines, and to demonstrate their own lack of liability in the matter—took issue with the testimony characterizing ATVs as unsafe. Howard P. Willens, a lawyer with Wilmer, Cutler, and Pickering, represented Honda, Kawasaki, Suzuki, and Yamaha before the Lieberman committee. Willens pointed out that an industry ATV task force had determined that nothing in three- or four-wheelers was "inherently defective." Many of the injuries associated with the machines had resulted from operator conduct strictly warned against by the manufacturers. Industry lawyers had made it clear to the Justice Department and the CPSC that the manufacturers were unwilling to sign a consent agreement committing them to remedies like the refund/repurchase program for three-wheeled ATVs that had been sold to children under sixteen. Such remedies, Willens said, were "not predicated on any statistical or engineering analyses, and . . . the industry would necessarily have to litigate any such proceeding that sought refund and repurchase remedies." But, he continued, industry leaders decided in the end to avoid lengthy, costly, and uncertain litigation. Magnanimously, and "in the public interest and those of the parties involved," the manufacturers therefore sought a settlement. The settlement involved very little cost to themselves. It amounted to an informational program and restrictions on sales to minors. As for recent reports on unscrupulous dealers selling adult vehicles to children, Willens disputed the validity of the surveys. The manufacturers believed that safety brochures and a ban on sales to youngsters would suffice to prevent deaths.[105]

In September 2002 the Consumer Federation of America and other groups petitioned the Consumer Product Safety Commission to go beyond voluntary standards. At hearings in the summer of 2003 in West Virginia and Alaska, dozens of parents, doctors, and public officials testified about the need for a ban on the sale of adult-size ATVs for use by children, while manufacturers' representatives and some users blamed the public health problem on operator misuse and lack of training. ATV-related injuries, according to best estimates, had doubled in a five-year period, and death rates continued to rise. ATV injuries requiring emergency room visits increased by 104 percent between 1997 and 2001, with a third of the victims in 2001 being under sixteen years of age. In 2005 the Consumer Federation reiterated its request to the CPSC to take action. In June of that year Hal Stratton, chairman of the CPSC, ordered the commission to review (again) the existing ATV safety standards, with the goal of making recommendations about the "advisability and potential for issues [of] an advance notice of proposed rulemaking."[106] Although having the authority to take direct action to protect the public interest, he followed this announcement in October with such an "advance notice of proposed rulemaking." He declined to ban adult-size four-wheel ATVs sold for the use of children under the age of sixteen, opting for further public comment. Stratton's decision is baffling, since the commissioners had noted 136,100 emergency room–treated ATV injuries in 2004, an increase of 8 percent over 2003 and 101 percent over 1998. Children accounted for 44,700 of those injuries, an increase of 16 percent over 2003 and 78 percent over 1998.[107]

What Is the Sound of One ATV Riding in the Forest?

Many ATV users and industry representatives claim that unnecessary regulation puts the industry at a disadvantage in competitive global markets. They point to efforts of the Environmental Protection Agency, the Consumer Product Safety Commission, and other federal agencies to require cleaner, safer vehicles at costs too high to conduct business, and that there is no proven need for cleaner, safer vehicles—only a need for more responsible operators. They worry about state and local restrictions on ATV access to public lands and about fees (taxes or registrations) that discourage buyers. Yet product improvements and growing sales indicate that ATVs are here to stay in large numbers—today there are at least seven million of them—and that manufacturers, owners, and other recreationists have seen the need to move toward rules and regulations that will enable safer operation and protection of public and private property. But this will require a shift in world view among operators and officials.

Forest Service chief Dale Bosworth, the nation's top forester, has called for a more coherent plan to regulate ATVs and other off-road vehicles.[108] A career specialist in forestry, Bosworth has nevertheless shown reluctance to regulate the machines that have ruined many boreal ecosystems. In a speech before the ATV Expo in Louisville, Kentucky, in October 2004, he welcomed the opportunity to work with the recreational machine industry to maintain access of ATVs and other off-road vehicles on federal lands. He cited the dubious honor of being the first Forest Service chief to attend the ATV Expo. He saw this not as a conflict of interest but as a sign of how far the Forest Service had come in fifty years of supervising a shift to motorized outdoor recreation. He proclaimed that off-highway vehicles "are a legitimate use in the right places. That includes many places on national forest lands." Hence the Forest Service would work to sustain that use for the future. He recognized that "relatively light" recreational use of forest lands in the 1960s meant minimal user impacts and conflicts. This had changed in recent decades with "tremendous improvements in OHV technology and tremendous growth in use." He acknowledged an "explosion" in use to nearly twelve million motorized visits annually, visits in which more than half of the users traveled more than 50 miles "just to have the opportunity to ride on national forest land, and about a third of them say they have no alternative—nowhere to go." Bosworth highlighted the fact that over 200,000 miles of forest roads (60% of the total) were open to use, supplemented by 36,000 miles of trails (or 28% of the total trail system).[109] ATVs simply cannot be avoided in national forests.

Bosworth recognized that this use had created a new sort of problem. Even if only 1 percent of users were irresponsible, they would generate unacceptable damage. Bosworth was aware of destroyed wetlands, riparian areas turned into mud, streams ruined, trails so deep "you can literally fall in," noise, and so on. In 2003 alone, users created more than 14,000 miles of illegal trails. Hence the Forest Service had proposed a new rule based on the assumption that off-road vehicles "are a legitimate way to enjoy the national forests and grasslands when they are used responsibly." The rule would not open or close a single trail or area but would allow local managers to decide the matter.[110] Yet thirty-five years of calling for responsible use and allowing managers at the local level to determine policy has meant insufficient study, no systematic regulation, and, as Bosworth himself noted, more and more use, often illegal use—in a word, continued irresponsible use.

Bosworth's position on ATVs provoked concerned among millions of Americans. In April 2004, representatives of over three hundred organizations wrote him to encourage reforms in management practices, including a more restrictive approach to safeguard national forests from ruin by ATVs. They worried about the "woefully inadequate" number of law enforcement officials in the Forest Service, where the typical ranger was responsible for more than 460,000 acres, an area more than half the size of Delaware. In spite of the Nixon and Carter executive orders, little had been done. They urged Bosworth to establish a two-year time line for implementing a plan, after which any forest that had not completed designations or closed trails would be permitted to allow ATV use only on roads that had been approved. The designations would be in any event in accordance with the National Environmental Policy Act and prohibit ATV use in wilderness-quality land.[111]

Several members of Congress—mostly from western states, all of them Republicans—welcomed the election of George W. Bush in 2000 as the most direct way to overturn years of policy aimed at safeguarding natural resources from machine recreation, policy that they believed gave too much consideration to the interests of nonmechanized recreationists. They wished to put the operators of ATVs and other recreational machines in the position of having to observe "responsible use" only as they themselves defined it.

These members of Congress spoke about the importance of the "country's natural resources" and the confidence "that we can manage our resources and public lands through good stewardship while maintaining their ecological integrity." Yet they believed that the costs of statutes and rules were greater than the benefits. They referred to the "the Clinton Administration's unreasoned and frequently absurd interpretation of law and Congressional intent" and the "misguided direction" of resource management that had "greatly limit[ed] access [of] the citizens of this country." (During the Clinton administration, the EPA and other federal agencies had begun to enforce clean air and water statutes more rigorously, to consider the growing public health menace of noise from recreational machines, and to restrict access to national lands, as stipulated in thirty years' worth of presidential and judicial orders issued by Republicans and Democrats alike.) By invoking "access," these western congressmen signaled that they intended to protect the rights of machine owners against the wishes of citizens who worried about the public health and environmental costs of recreational machines. They proposed to keep parks open to snowmobile use, overriding long-term studies that led to the inescapable conclusion that snowmobiles should be phased out. They rejected resource preservation as required by the 1916 Organic Act, instead voicing a preference for "visitor enjoyment." They opposed the practice of designating lands as wilderness or national monuments under the 1906 Antiquities Act.[112]

Government officials have tried to permit fair use of ATVs and other off-road vehicles on federal lands. Their responsibility is to balance the use of land and the resources on it among competing interests. Unfortunately, fair use of ATVs has been nearly impossible to achieve. There are too many machines, and their impact on ecosystems is immediate and enduring. Where ATVs tread, their noise, smoke, and environmental degradation have reduced recreation for hikers to a dream. All too often regulators have sided with manufacturers to keep lands open. They frame this as ensuring the rights of machine owners. They assert that "fair use" requires that access to lands be maintained for these individuals. They argue that voluntary regulation by manufacturers and clubs is sufficient to ensure responsible use. And they say you must not restrict use just because a few irresponsible operators go off the trails, ride while intoxicated, or hurt themselves or others in accidents.

Yet remember: one in three ATVs will be in an accident in its lifetime. This is not a matter of the occasional bad apple ruining it for others, the irresponsible or drunk operator plowing through a stream, the troublemaker revving the engine to scare animals. By their very nature, ATVs will almost inevitably destroy wilderness, forests, grasslands, and dunes. Their wide and long wheelbase facilitates access everywhere; their tires crush and churn, making all trails wider; their engines pollute. Too many riders are hurt or killed even though they drive as instructed. The ATV industry claims that, beyond educational programs, no further measures to improve safety are necessary. Yet ATVs are inherently unstable owing to a high center of gravity, and their users drive them at high speeds over difficult terrain because that's what they're for. Will the "dumbing down" of the ATV be the only way to ensure its safety?

4

THE LUNACY OF PERSONAL WATERCRAFT

THE LOON IS BELOVED among campers, canoeists, and other recreationists for its beauty, its haunting cry, and its extraordinary skill as a swimmer and diver. For breeding, loons prefer lakes larger than 60 acres with clear water and an abundance of small fish. They nest close to the water's edge, on small islands, irregular shorelines, floating bog mats, and marshy hummocks. The Migratory Bird Act of 1918 affords protection to the loon even though it has not been formally listed under the federal Endangered Species Act.[1] Yet the loon is threatened by human pressures on its habitat, pressures like the construction of extravagant summer homes, the spread of such persistent organic compounds as chlorinated pesticides, and the unintentional intrusion upon nesting areas by doting recreationists. Evidence of a decline in loon reproductive capability has been accumulating since the 1950s. Massive die-offs of loons occurred in Lake Michigan in the falls of 1963, 1964, and 1965. Wildlife biologists worry about the increasing numbers of canoeists and operators of motorized boats who explore loon habitats. Many of these individuals make their annual pilgrimage to loon country in the spring, during fishing season, precisely when nesting occurs. After frying and eating their catch, some of them unthinkingly throw the fish entrails, bones, and other scraps into thickets near the nests. Ravens descend to eat the remains of the meal.[2] The frightened loons, their habitat already under pressure from increasing numbers of summer cottages, move off, occasionally leaving broods to fend for themselves. And now the personal watercraft (PWC) has entered into this confluence of construction, chemicals, and recreation.

Loons attract public interest and concern. They adorn a special license plate of the state of Maine; the extra cost of the license plate generates funding for conservation programs. A large number of nongovernmental organizations (NGOs) dedicate their activities to protecting the loon. They recognize that in preserving the loon, we preserve northern aquatic wilderness. As biologist David Evers notes, loons serve as indicators of aquatic health in decline and human encroachment on habitat, no matter how resilient they may appear to be. Loon breeding populations are restricted to four countries (Iceland, Greenland, Canada, and the United States), with their winter range extending to both the Atlantic and Pacific coasts and the Gulf of Mexico including the Florida Keys. The breeding range covers the lakes of Canada and sections of northern Minnesota, Wisconsin, Michigan, and New England.[3] PWCs have assumed the same habitat and range.

Any disturbance of these regions, seasonally or otherwise, runs the risk of lowering loon reproduction rates, threatening chicks, and reducing population. Citing evidence that loons have been in decline for decades, PWC operators claim that their machines have no direct link to the decline. Yet the evidence of the machines' impact—at first anecdotal, then increasingly quantitative—resembles the evidence of the threat other motorized craft pose to nesting birds. Like other machines, PWCs destroy habitat by emitting noise and pollutants and by churning up wakes. They may be even more insidious than other recreational machines because they can approach nesting areas closely by virtue of their shallow draft.

PWC operators prefer to use them in two kinds of ecosystems: coastal wetlands and lakes. Coastal wetlands are among the most productive and ecologically valuable of the world's ecosystems. They have been prominent in such cradles of civilization as the coastal estuaries of the Middle East, China, and India. Marine-dependent settlements date from the second millennium B.C. in North America. Wetlands are also among the world's most heavily disturbed ecosystems. Because they are productive and serve as transportation arteries, they attract human settlement. More than half of the U.S. population live within 50 miles of the coast, and perhaps 70 percent of all humanity live in coastal zones. The impacts of human occupation include alteration of hydrological processes, introduction of pesticides, extensive agricultural run-off, nutrients, heat, and exotic species, and unsustainable harvest of native species. Perhaps only 46 percent of original wetlands in the United States remain.[4] Gradual, persistent, and irreversible destruction may not instantly be recognized. We usually admit the importance and fragility of wetlands only during disasters, for example the tsunamis that struck eastern and southern Asia in 2004 and the hurricanes, Katrina and Rita, that devastated the Gulf coast from Texas to Florida in 2005. But increasing numbers of recreational boats and jet skis have accelerated this degradation. Accidents frequently result. Public health and environmental officials have had no choice but increasingly to regulate PWC and other boat operation and to restrict the areas where they may be used. Additional pressure from recreational vehicles must be considered in this light.

Like any fast and thrilling vehicle, the PWC is dangerous. Operators face the risk of blunt trauma injury and death, as do the swimmers and birds who happen into their path. They do not require driver's licenses to operate in spite of the fact that they have a quirky if not inherently dangerous design flaw: they can be steered carefully only when the throttle is engaged. Few states require operators to undergo special training. They can be maddeningly loud, not only because of the high-decibel whines the engines emit but also because they slap the surface of the water, constantly changing the tenor and intensity of the sound. While the engines in new PWCs meet Environmental Protection Agency and California Air Resources Board standards, most PWCs still in operation use older, two-stroke engines that funnel pollutants into the air and water. Finally, like other machines, once PWCs have appeared in any recreational area, they are difficult to restrict. There are 1.5 million PWCs in the United States, and their numbers are growing, with sales reaching a peak of 200,000 units in 1995. Sales have fallen significantly since the mid-1990s, perhaps because of the quality control problems that have plagued the industry from the start; roughly one-fifth of all the vehicles manufactured have been recalled.

According to the U.S. Coast Guard, PWCs are inboard boats under 16 feet in length with a two- or four-stroke engine that drives a jet pump. The pump draws water from the bottom of the craft into an impeller (a type of propeller fitted into a surrounding "tunnel") that pressurizes the water and forces it out a nozzle at the rear of the craft. The jet of pressurized water propels and steers the craft when the throttle is engaged. Two-person craft have become more popular than the single-person models, and today three- and four-person family models are showing the strongest growth. Multiple-person family craft make up more than 97 percent of current sales.[5]

Recreational vehicles found a niche in postwar America because of consumers' increasing disposable incomes and the growing number of artificial recreational areas. As part of their flood control efforts and attempts to improve upon the water regimes of various bodies of inland water, the Army Corps of Engineers straightened rivers and shoals, dredged shallow areas, and built various impoundments. In cooperation with such engines of hydrological change as the Bureau of Reclamation, the Tennessee Valley Authority, and the Bonneville Power Administration, the corps changed the face of the Mississippi, Columbia, Tennessee, and dozens of other river basins. They created barriers large and small for flood control, hydroelectricity, or both. The lakes formed by the projects were a boon for inland fisheries and recreational purposes, although sports and commercial fishing has fared poorly in terms of high costs, low productivity, and environmental ramifications.[6] Not so for recreation: the pleasure boat industry grew rapidly to fill the waters. Through the 1970s these were powerboats, speedboats, and motorboats. In the 1970s PWCs were added to the mix of pleasure, recreation, and thrill on inland waterways.

Several powerful government bureaucracies—the Fish and

Wildlife Service, the National Park Service, the Environmental Protection Agency, and the National Oceanic and Atmospheric Administration (NOAA)—have joined the Corps in considering the place of nature among the interests of all groups. But they often work at the opposing purposes of environmental protection and economic development. They claim to protect biodiversity while encouraging large engineering projects, particularly those that draw freshwater out of wetlands. For Florida waters the situation is especially confusing because state and federal agencies each have a different take on the place of PWCs in coastal waters, and disputes exist even among state or federal agencies. PWCs have been banned from sixty-six of the eighty-six national parks, from Lake Tahoe in California, and from Washington State's San Juan Islands.[7] But they persist and multiply wherever they are not prohibited, and the administration of George W. Bush has actively sought to scale back or postpone restrictions. Recognizing growing public concern about PWC noise and pollution, manufacturers have introduced cleaner, quieter, and more fuel-efficient models. Still, loons and other magnificent birds may soon vanish from the areas in which PWC operators love to cruise because of all the pressures on habitat, of which PWCs are a major new source.

The Call of Nature

Like other recreational vehicles, personal watercraft allow operators to move quickly—but not quietly—through various ecosystems. PWCs (often called jet skis after the Kawasaki model) can reach top speeds of 90 mph. They skim and jump over the surfaces of lakes, rivers, ponds, and oceans. Today there are four major manufacturers of PWCs: Kawasaki (Jet Ski), Yamaha (Wave Runner), Bombardier (Sea-Doo), and Honda (AquaTrax). These four manufacturers dominate the market, producing together up to thirty different models annually. (Arctic Cat manufactured the Tiger Shark for about ten years but left the business in 1999, and Polaris Industries left the market in 2004.) The 1998 models ranged in cost from $4,800 to $9,400, constituted more than 36 percent of all new recreational boats sold, and represented more than $1.2 billion in annual sales. PWCs in 2005 cost $10,000 to $12,000. Other necessary expenditures include $100 to $200 for a wetsuit, $100 to $150 for a life preserver (personal flotation device, or PFD), $70 to $90 for gloves and booties, and $200 for helmet and goggles. The trailer and hitch might run another $1,500.

Contemporary PWC engines generally range from 62 hp to 135 hp, with displacement ranging from 639 cc (with two cylinders) to 1,131 cc (with two or three). By comparison, the first Jet Ski was rated at 32 hp. The largest PWCs have 1,300-cc engines and power in at 430 hp, comparable in size to a small automobile engine and exceeding many of them in power. At this size and power, PWCs no longer count as small-bore engine vehicles. PWCs have an average operating weight, with gasoline and rider, of roughly 700 pounds.[8] Whereas a traditional boat may have a horsepower-to-length ratio of 4:1 (e.g., a 65-hp engine on a 16-foot craft), in PWCs the ratio may be as high as 12:1. These machines are made for speed and play. Snowmobiles may permit people in isolated communities to maintain contact in winter, ATVs may allow law enforcement officials to help stranded hikers, but PWCs have one purpose: to go fast while thrilling their operators (although manufacturers have made them available to wardens, rangers, and law enforcement personnel for utility purposes).

From Standup to Sit-Down: The Evolution of the Personal Watercraft

Most observers credit Clayton Jacobson II of California (or, according to some accounts, Arizona), a motocross enthusiast, with the invention of the standup model PWC (he received a patent in 1971). Even earlier, in 1955, the Vincent Motorcycle Company had marketed the propeller-driven 200-cc Amanda Water Scooter. According to motorcycle enthusiast Mike Nixon, the PWC was introduced to the public in 1977 in a major motion picture, the James Bond movie *The Spy Who Loved Me,* in which "actor Roger Moore rode a water-going motorcycle, a one-off, Tyler Nelson–patented, ski-steering craft which was later manufactured by Spirit Marine, then a division of Arctic Cat."[9]

The first commercially successful PWC, the Jet Ski, was introduced by Kawasaki in 1972. As reported in the *San Diego Union* in October 1972, people saw a watercraft "skimming the water behind a pair of motorcycle-type handlebars testing a new type of water scooter" that gave the "thrills of water skiing without the need of a towboat."[10] Kawasaki turned to mass production of the Jet Ski 400 in 1975 at its Lincoln, Nebraska, plant. Unexpectedly strong market demand for more power led to the development of the JS440 in 1977. The next improvement was the JS550, introduced in 1982 with a high-capacity mixed-flow pump, driven by a water-cooled 531-cc two-stroke engine.[11]

Kawasaki models from the late 1970s and 1980s were designed for one-person, standup operation. They were tiring to operate and difficult to handle. During the late 1970s manufacturers introduced the sit-down style that has become the industry standard.[12]

The history of Kawasaki's Jet Skis parallels that of the industry as a whole, with uninterrupted growth in the power and size of engines. In the mid-1980s Kawasaki focused simultaneously on comfort, convenience, and family boating and on building more powerful performance machines with larger and larger engines. In 1986 the company appealed to new riders with the lightweight and easy-to-operate JS300, which had a single-cylinder 294-cc engine with automatic fuel and oil mixing. It added a hybrid two-passenger model with standup and sit-down capabilities (the X-2) with a water-cooled, 635-cc two-stroke twin-cylinder engine developing 52 hp at 6,000 rpm. The water-jacketed exhaust system helped moderate noise levels. For the sports crowd, Kawasaki introduced two standup models, the "wet and wild 300SX" and the 650SX, a powerful model with a new V-hull design that increased stability during boarding and high-speed maneuvering. The two-seater Tandem Sport, with a 635-cc two-cylinder engine, was the company's first true sit-down PWC.[13]

Early models challenged even expert riders. One journalist noted in 1978 that he was rarely able to stay on a Jet Ski for more than ten or fifteen seconds, although he enjoyed the "dizzying 20 mph." The Jet Ski "retains the stubborn attitude of an unbroken bucking horse until taught that *it* is the drivee. This, however, is not achieved without effort. . . . Once the initial threshold is crossed, the Jet Ski is really quite friendly, likable . . . and above all safe. I kept looking for defects and couldn't find any." He also noted that the machine's 243 pounds felt heavier and heavier as the day wore on. There's no problem if you fall off, because the throttle idles and the Jet Ski "circles gently, waiting to be reclaimed." Contrary to what we have heard, he claimed, "There's no way little Johnny will come screaming back to the cabin trailing blood from what used to be his left arm."[14]

High-revving, high-performance vehicles dominated the sales line in the 1990s, including the 550SX in various versions and the more powerful "upgraded" 650SX with a "a high-revving, in-line, twin-cylinder engine" that "satisfied the incessant craving of racers for more power." The 650SX featured a "new underwater exhaust outlet [that] reduced noise levels while allowing peak performance." In 1992 the company introduced the 750SS two-seater with a twin-cylinder 744-cc engine "fed by a huge 40 mm carb mounted to 8-petal reed blocks" that gave the machine "a massive spread of responsive power. Rubber engine mounts improved comfort and reduced vibration stress on the rest of the boat. The tough, fiberglass-reinforced hull was fitted with storage compartments and a large-capacity fuel tank for long-range fun." Kawasaki also introduced dual-carburetor models like the Super Sport Xi, which executives likened to a "two-seater hot rod." It was followed in 1994 by the Super Sport XiR and the ST, Kawasaki's first three-seater. For sheer raw power, the company offered the 900 ZXi, its first three-cylinder model, with an 891-cc crankcase reed-valve engine. It came with adjustable rear-view mirrors.[15]

But even this was not enough power. In 1996 Kawasaki introduced the 1100 ZXi, "powered by a bored out version of the 900 ZXi's 3-cylinder engine." The engine had a displacement of 1,071 cc and developed 120 hp at 6,750 rpm. Needless to say, the 1100 ZXi was the most powerful PWC on the market. It had three carburetors "fitted with accelerator pumps for instant throttle response and instant acceleration." Then came the Ultra 150, with close to 150 hp and a 1,176-cc engine. It was equipped with Kawasaki's "Throttle Responsive Ignition Control," which "continuously altered the timing of the digital ignition for each individual cylinder to suit operating conditions."[16] The horsepower wars were in full swing.

In the first decade of the twenty-first century, in response to increasing pressures from state governments to make PWCs safer, quieter, and less damaging to ecosystems, Kawasaki and the other manufacturers began to stress "green" engines that met the 1996 EPA standards. PWCs were increasingly equipped with direct-injection two-stroke engines to curb emissions and reduce fuel and oil consumption. When the 2006 EPA standards were introduced, emissions of the direct-injection engine were already low enough to meet them. Manufacturers also began to address the problem of off-throttle steering. Kawasaki introduced its trademarked Smart Steering system, a system that helps the rider turn at running speed, even without the throttle being applied. In a departure from the greenness and safety of these new models, Kawasaki also introduced "Ninja performance on the water," the four-stroke STX-12F with a four-cylinder fuel-injected 1,199-cc engine.[17]

In 1968 Clayton Jacobson, the inventor mentioned earlier, collaborated with Bombardier to develop a wide, flat-bottomed, aluminum-hulled sit-down PWC with a 318-cc Rotax engine: the Sea-Doo, a marine counterpart to the Ski-Doo. The 24-hp 30-mph Model 320 was "slow steering and rough riding." In 1969 Bombardier introduced a water-cooled 368-cc engine in its Model 372. This machine reached 35 mph, but while performance and reliability had improved, making corrosion-resistant engines and parts for constant exposure to salt water remained a challenge.[18] At this point Laurent Beaudoin, the company's current chairman and chief executive officer, was its thirty-year-old president. Following the example of snowmobile trailblazers who had pushed sales in the first years with "snofari" trips through Alaska, "snodeos," and races, Beaudoin and Sea-Doo project designer Anselme (Sam) Lapointe joined ten other people on the first long-distance Sea-Doo tour. The four-day 469-mile cruise from Montreal to New York City highlighted the many unique attributes of the craft: stability, versatility, freedom, and fun. The Sea-Doo was a risky venture for Bombardier in the first place, given its radical design compared with traditional powerboats and the need to meet the rapidly growing snowmobile demand that was the company's bread and butter. Though they needed a summer product to keep factories open, the managers were so uncertain about the Sea-Doo's future—and so concerned about the unreliability of the Rotax engines owing to corrosion—that they shelved the vehicle in 1970. Only in 1986 did the company reevaluate the market situation and reembark on the manufacture of Sea-Doos.[19]

By the early twenty-first century, Bombardier's Sea-Doo had become the leading PWC on the market. In 2003 the Bombardier Sea-Doo accounted for nearly one of every two PWCs sold. Current annual sales in North America, Asia, Europe, and Africa are roughly 200,000 units. Today's "fully marinized Rotax engines, corrosion-resistant parts and scientifically designed hulls" have solved the earlier problems. Company representatives credit their success to a number of design innovations. Bombardier introduced the first three-passenger watercraft in 1990, the first high-performance watercraft in 1991, the first Runabout National Champion and the first two-time "Watercraft of the Year" in 1992, the first industry "Best Buy" in 1993, the first watercraft suspension in 1995, the first computerized theft deterrent system in 1996, the first mass-produced fuel-injected watercraft in 1998, the first watercraft line to incorporate significant sound reduction in 1999, and finally, in 2000, the first "Watercraft of the Century" winner, the first "Watercraft of the Year" winner for the new millennium, and the first marine manufacturer to win the IDSA "Designs of the Decade" medal. In 2004 it introduced the first watercraft to surpass 200 hp, the extravagant 215-hp Sea-Doo RXP, powered by an engine that "hums with aggressive innovation." If the RXP isn't powerful and gas-chewing enough for you, newer Sea-Doos "propel you into the future of fun on the water with state-of-the-art power packs such as the 430 horsepower of the twin Rotax 4-TEC engines in the Speedster 200 and the new Speedster Wake."[20]

On a recent visit to the new state-of-the-art Sea-Doo assembly line at BRP (Bombardier's recreational equipment spin-off), I observed contemporary assembly practices geared toward building PWCs to the highest specifications and tolerances. Bombardier takes advantage of team assembly practices common at such automobile manufacturers as Toyota. The facility in Valcourt is spacious, well lit, and virtually free of dust and odors. Robots assist workers in building the machines. For example, a "spider" performs the clamping necessary to glue the hull to the top in a unibody. Three-cylinder Rotax four-stroke engines meet all emissions standards. The workers are not unionized, a fact the managers attribute to the excellent and safe working conditions and the high pay.

Controversies and Uncertainties

PWCs are both extremely popular and of great economic significance. Approximately twenty million Americans ride them each year. PWC manufacturers employ six thousand people in at least eleven states. There are more than two thousand retail businesses servicing and selling the machines, and any number of aftermarket and other related small businesses.[21] Yet the large and increasing numbers of Sea-Doos, Jet Skis, AquaTrax, and Wave Runners that have entered fragile ecosystems have raised serious concerns among many observers about whether and in what ways they ought to be restricted.

Florida's nearly 1,200 miles of coastline and its many thousands of miles of tidal shore land and beaches have attracted the largest number of PWCs registered in any state. Registrations of PWCs are growing at a rate of 2 percent annually and will soon reach one million. Even with all that coastline, Florida boat traffic has become heavy. Granted, other activities and machines also threaten Florida's natural beauty, but this in no way lessens PWC's contribution to the problem. Florida has 1,263 federally listed threatened and endangered species. Yet shoestring budgets, based almost entirely on user fees (licenses and excise taxes on sporting goods) have left state fish and wildlife agency programs underfunded. Florida's citizens, and its flora and fauna, must rely on the work of volunteers and foundations to provide any hope of maintaining the integrity of state programs for preservation of ecosystems.[22] While some PWC clubs assist in ecosystem protection activities, most of them focus on maintaining access to areas already under threat, and many of them have been reluctant to acknowledge the environmental costs of their activities.

The difficulties of regulating PWCs in Florida stem in part from its "antidiscrimination law." Florida is the only state with a law prohibiting discrimination against any particular type of boat: what holds for a motorboat must hold for a PWC. But PWC users feel that they really do face discrimination compared with other boat users. For example, in Biscayne Bay National Park—the largest marine park in the national system—PWCs are prohibited while boats are not. Everglades National Park—the only subtropical preserve in North America containing both temperate and tropical plant communities, including sawgrass prairies, mangrove and cypress swamps, pinelands, and hardwood hummocks, as well as marine and estuarine environments—prohibits PWCs but allows other boats (although with certain exceptions and restrictions). The Florida Keys National Marine Sanctuary is managed not by the National Park Service but by the National Oceanographic and Atmospheric Administration. The sanctuary prohibits PWCs except in four designated areas, whose access routes are regulated by both NOAA and the state of Florida. And even more confusing, the National Park Service prevents PWCs from using the Intracoastal Waterway that runs from Miami to the Florida Keys; for safety reasons, they must remain 15 miles outside the waterway, while barges, ships, tugs, and yachts may navigate through it. Peggy Mathews, formerly a Florida state employee, then with the Personal Watercraft Industry Association (PWIA) and now the Florida representative of the American Watercraft Association (AWA), said, "The spectrum of regulations is simply so confusing."[23]

What is the basis of "discrimination" against PWC operators? Wildlife conservationists have long recognized the impact of recreation and ecotourism on wildlife. Declining reproductive success, species diversity, and density indicate this impact. With PWCs the sound, fury, and extent of the impact became much more proximate. Several researchers have tried to determine if buffer zones can be established to limit human disturbances near critical areas. In one study, scientists with the Bureau of Wildlife Diversity Conservation of Florida exposed twenty-three species of waterbirds to PWCs and outboard-powered boats to determine flush distances. They recommend buffer zones for both boats and PWCs of 100 meters for plovers and sandpipers to 180 meters for wading birds.[24] Subsequent research showed that larger species generally exhibited greater average flush distances when exposed to fast-moving outboard-powered boats or airboats. After a series of measurements conducted from 1999 to 2000 to identify minimum buffer zones for foraging and loafing waterbirds, the researchers recommended that buffer zones be extended significantly, from 130 meters to 365 meters for such raptors as the bald eagle, to 165 meters for the tricolored heron, and to 255 meters for the great egret.[25]

PWC operators believe that since both boats and PWCs flush birds, then both should be allowed, and that since PWCs arrived on the scene only recently, they cannot be held responsible for the long-term decline of various species. As noted, they argue that bird populations have been in decline since the 1940s and 1950s, especially fish-eating species, including the common loon. Yet to point to historical trends or to the impact of other human activities does not change the fact that PWCs have a harmful effect on the bird populations of the lakes and estuaries they frequent.[26] Some of that harm is intentional. Because loons can swim underwater for long distances, many people chase them to see where they will surface. Kayakers, canoers, and, increasingly, jet-skiers have pursued the loons to exhaustion. In 1997 there were three confirmed incidents of loons being hit and killed by speeding watercraft on northern Michigan lakes. In Otsego County two jet ski operators "literally terrorized the loons right out of existence."[27] The Audubon Society investigated the reports of the Michigan Loon Preservation Association that PWCs had chased or hit birds. In 1998, when only four hundred nesting pairs remained in Michigan, a state where thousands of loons once nested, there were thirty-two significant incidents of loon harassment, up from only one incident in 1986 when the PWC was first introduced.[28]

Some loon aficionados believe that "loons and jet-skis can peacefully co-exist on our northern lakes," and that jet ski operators can learn to understand "the consequences of their noise and wake, stay out of nesting areas, slow down and never chase the loons." If "harassment continues or becomes intentional," observers must contact rangers, wardens, or police and must document the offense with photos or videotapes to have any chance of successful prosecution.[29] Indeed, local, regional, state, and national officials now confront the need to set aside lands for habitat and to restrict access of such recreational machines as PWCs precisely because many operators do not recognize that their riding behaviors threaten habitat.[30] Small steps toward restriction, rather than outright prohibitions in some areas, may be insufficient, given the fact that throughout North America people are taking more and more "wilderness vacations." They pursue hiking, canoeing, kayaking, and so on in what they perceive to be the "wild." Increasingly, they engage in the motorized versions of recreational activity—motorboating, ATVing, snowmobiling, and jet-skiing—all of which have a far greater, more immediate, and more lasting impact than nonmotorized activity.[31]

Is the evidence that PWCs destroy habitat and threaten wildlife incontrovertible, or is it subject to dispute? Birds, fish, and invertebrate species have suffered significantly from all kinds of recreational boating. The impacts are most pronounced in shallow-water areas, where many species nest and feed. The problems are compounded by the fact that peak boating times coincide with critical life stages of the species. Outboard motorboats and PWCs generate tremendous engine wash, which can damage benthic eggs and larvae. Shear and rotational forces destroy fragile organisms. Death also occurs when organisms are smothered or buried by sediments. Pollutant emissions hurt marine creatures at every stage of life, beginning with egg mortality. (See table A.5.)

Granted, most wildlife disturbance is due to inappropriate or irresponsible operator behavior compounding the effects of the machine itself. The jury is still out on whether the effects of PWCs are greater than those of other vessels. Yet whether they are less, greater, or the same, those effects occur, they are damaging, and they often provoke irreversible damage of habitat, increased mortalities, or decreases in mating, and they occur for all machines, responsible operator or not.[32]

Representatives of the Personal Watercraft Industry Association dispute the contention that PWCs and nature are somehow in conflict. They cite a series of reports to argue that PWCs have no greater impact on waterfowl, and perhaps even less, than other motorized boats. They argue that "personal watercraft are jet-powered, they have minimal impact on seagrasses, marine mammals, fish and other aquatic life. Although the small draft of a PWC allows it to operate in shallower water than other boats, PWC users never intentionally operate in grassy or extremely shallow areas as the grass or other debris can cause serious engine damage when sucked into the engine, leaving the operator stranded and with an expensive repair bill."[33]

Commenting on the studies by the Florida Fish and Wildlife Conservation Commission comparing flush distances of PWCs and two-stroke engine motorboats, PWIA representatives claim that the researchers found PWCs to be "relatively quiet to the point where their noise is not the factor which causes the birds to flush. . . . A fast moving motorboat heading directly at the birds . . . should produce a flushing response similar to that of a PWC being operated in a similar manner." Greater flushing distance for the PWC over the motorboat existed in only one species, while for five species the flushing distances were greater for motorboats. PWIA officials noted that the researchers also suggested a single buffer zone for all watercraft to protect nesting waterfowl, not a special, more draconian zone for PWCs. They site another Florida study that found an average greater flush distance of birds in response to people walking than to approaching motorboats or canoes.[34]

Peggy Mathews, Florida AWA representative, proudly contends that PWCs "are [the] cleanest, quietest boats on the water." Mathews referred to the research of James Rodgers and others, which found PWCs to be no more intrusive than other boats with respect to wildlife. She dismissed research that demonstrated otherwise as "anecdotal and mere observation." To call these studies "peer reviewed—that's crazy," she said. Mathews also referred to several studies—some commissioned by the PWIA itself—that found PWCs to be safe to aquatic life. For example, a study conducted by Continental Shelf Associates indicated that PWCs have no impact on benthic organisms like seagrass if used according to manufacturer's recommendations even in water as shallow as 24 inches.[35] It should be noted that Continental Shelf Associates is a group of "industry professionals" who serve the oil and gas, telecom, government, and scientific markets,[36] a fact that also raises the issue of peer review.

In many ways the PWIA has taken a selective approach to research on the effects of PWCs, embracing any results that support their desire to avoid restrictions on the machines and discounting the rest. For example, Florida studies conclude that human disturbances affect an animal's ability to feed, breed, and rest and can disrupt wildlife community dynamics—not that PWCs have no impact. Since even walking and jogging can disturb waterbirds near shorelines, beaches, sandbars, and islands, researchers frequently call for restricting those activities too. The Florida Fish and Wildlife Conservation Commission researchers called not for eased restrictions on PWCs—even though there was considerable variation between the flush distances of PWCs and motorboats—but for a single, and relatively large, buffer zone for both kinds of craft.[37] Still, on the basis of this research, PWIA representatives concluded that species type is more important than boat type when determining boundaries that should not be crossed by humans. Should the conclusion logically be that PWCs ought to operate where less valuable species nest? That PWCs are not disruptive to ecosystems? That we ought to replace motorboats with PWCs since the former flush some birds with greater frequency?

PWIA officials hope to avoid restrictions and prohibitions through voluntary measures. Will voluntary efforts to protect fauna and habitat from PWCs and other recreational vehicle disturbances suffice, or are prohibitions required to stem their assault on ecosystems? According to a 1995 survey conducted by the U.S. Fish and Wildlife Service, recreational powerboat activity occurred on nearly one-quarter (117 of 504, or 23%) of the refuges within the U.S. National Wildlife Refuge System, while almost one-third of those refuges lacked the authority to regulate and control powerboat use.[38] Wildlife specialist Kevin Kenow examined whether voluntary restrictions in one refuge, the Upper Mississippi River National Wildlife and Fish Refuge, a congressionally authorized refuge dating to 1924, had worked to protect the refuge. Because it also comprises part of a commercial navigation system managed since 1930 by the Army Corps of Engineers, refuge officials permit both commercial and recreational boating in it.[39] Kenow noted that recreational boating disrupted feeding activities of diving ducks on Lake Onalaska, an impounded area of the Upper Mississippi River, which may lessen its quality as a staging site for the birds. Because of the importance of the area for these waterfowl, beginning in 1986, with the assistance of the Army Corps of Engineers, refuge scientists initiated a voluntary public program to restrict commercial and recreational boating activities. Then, in 1997, the Upper Mississippi refuge was designated a Globally Important Bird Area because of its critical importance in supporting global populations of the bald eagle, tundra swan, canvasback ducks, and other waterfowl—tens of thousands of which pass through the Mississippi Flyway.[40]

Scientists tried to communicate the biological importance of the lake through an extensive educational program. They distributed leaflets, established personal contacts, prepared kiosk displays, wrote letters to property owners, and made public service announcements. They closed approximately half of a roughly 7,200-acre area to hunting and trapping during the duck-hunting season.[41] The program was a success. Although boating traffic on the lake increased year by year, there was a decline in the proportion of boats that intruded into the restricted area. This means, of course, that in terms of absolute numbers, the intrusions increased. And pressures from boaters engaged in recreational and commercial fishing, government researchers, and hunters visiting barrier islands adjacent to the western boundary of the area remained high.[42]

Yet these voluntary regulations proved to be too much for many operators and manufacturers to tolerate. American Watercraft Association personnel adamantly opposed reductions in riding areas open to PWCs when the U.S. Fish and Wildlife Service proposed a new conservation plan for the Upper Mississippi Refuge. The plan offered four alternatives, three of which PWC owners claimed limited the way "taxpaying PWC owners would be allowed to access these public waters for the next 15 years."[43] But the proposals gave significantly less weight to other taxpayers—those who did not own PWCs, the vast majority of citizens—and their rights to the creation of a protected refuge without motorboats or PWCs. Fish and Wildlife Service personnel preferred a plan that permanently closed roughly 14,500 acres of water, or only 6 percent of the total, currently open to PWCs and other motorized craft. Still, industry officials characterized this modest plan as "the clear route to the slippery slope of a total prohibition of PWC in the refuge."[44] The real problem was that having allowed access to even one PWC led down the slippery slope to scores of them and to the great difficulty of regulating them retroactively.

Smoke, Noise, and PWCs

Although current models are much cleaner and quieter than those manufactured through the late 1990s, PWCs have engendered opposition in large part because of their noisy and highly polluting engines. Manufacturers acknowledged their ability to produce quieter, less polluting engines but waited to do so until forced by public reaction and regulatory pressures. As with snowmobiles and ATVs, the two-stroke engines powering PWCs were lighter, more responsive and powerful, and cheaper to manufacture than four-stroke engines. In 2003 all PWC manufacturers began to offer models with more efficient four-stroke engines, having belatedly recognized that "four-stroke marine engines are . . . one of the 'greenest' engine types on the water today." While they are somewhat more expensive than two-stroke engines, they are 75 percent cleaner than conventional two-stroke engines and could save consumers hundreds of dollars annually in fuel and oil costs. Manufacturers have also begun to produce direct-injection two-stroke and electronic-fuel-injected (EFI) plus catalyst two-stroke engines, which occasionally rival four-stroke models for reduced emissions. As Peggy Mathews told me, "Because of the technological advances of the last six years, we have quieted down our engines. EPA standards are met for 2006."[45] "We welcome scrutiny of our technology," said Kirsten Rowe, then executive director of the Personal Watercraft Industry Association. "Makers of personal watercraft have long been committed to change," she explained. "They have directed their resources to create some of the cleanest and quietest motorized boats on the water today."[46]

The direct-injection designs used by manufacturers since 1999 scavenge engine cylinders with pure air containing no fuel at all. The fuel is then directly injected into the cylinder after the exhaust port is closed. This not only results in significantly reduced hydrocarbon emissions but also improves fuel economy. Ficht and Orbital developed direct-injection systems and added catalytic converters to two-stroke engines to reduce hydrocarbon emissions by up to 80 percent. The fuel-injection technology of the Ficht engine provides improved fuel economy, quicker starts, virtually no smoke, and improved throttle response, and industry representatives considered it one of the best engines available. Still, these engines produce more noxious emissions than four-stroke engines.[47]

While the Ficht and Orbital fuel-injected two-stroke engines cut emissions substantially compared with older two-stroke models, they had their own set of troubles. Initially, severe production problems bedeviled the entire product line, generally in the form of fuel leaks that could cause fires and explosions.[48] Indeed, tens of thousands of PWCs were recalled because of the danger of fire, and several states now require PWCs to carry fire extinguishers. Bombardier, manufacturer of the Sea-Doo, had purchased the Outboard Marine Corporation, the maker of the Ficht engine, during the latter's bankruptcy proceedings. It sought "higher horsepower, greater efficiency and cleaner emissions, not to mention lower prices, increased durability and better customer service." Bombardier discovered problems in the engines and in May 2001, after a Coast Guard "public safety advisory" issued in March, agreed to recall over eleven thousand defective 200- and 225-hp engines built in the 1999 and 2000 model years by the now-defunct Outboard Marine Corporation, "to restore public confidence in the brand, particularly those with Ficht fuel injection technology."[49] Bombardier "made a number of changes to the Ficht engines to eliminate past problems. For example, it changed the stainless-steel injection ram ball to ceramic and has completely altered the electronic mapping to avoid carbon-loading at slow (trolling) speeds. All fuel connectors and hoses have been upgraded, and a vapor separator has been installed as a blow-off system to prevent any fuel leakage in over-pressure situations. The new Ficht engines handle poor-quality fuel and alcohol additives much better."[50]

EPA officials point out that the new engine technology is not only more environmentally friendly but also promises savings in fuel costs to offset the slightly higher cost of the new engines themselves. The new designs would also "relieve boaters from the hassle of mixing fuel and oil." Yet the EPA rules are not retroactive, and many years will pass before the older engines are retired, meaning that the EPA can expect to achieve a 50 percent reduction in emissions from the introduction of these new engines only by 2020, or perhaps later.[51] These older two-stroke engines may discharge 30 percent of their fuel unburned directly into the water; with some machines having consumption rates of up to 10 gallons per hour, up to 3 gallons of fuel-oil mixture may be released into the water for each hour of operation. According to the California Air Resources Board, emissions of hydrocarbons and nitrogen oxide from PWC and outboard engines in California alone amounted to approximately 777 tons per weekend day, or 50 percent more than the exhaust coming from all passenger cars on all California roads on a typical weekend.[52] Kayaks, canoes, and other conveyances can reach shallow waters where they may disturb wildlife, but they do not contribute to hydrocarbon pollution.[53]

As they have done in other pollution control efforts, officials in

California set the pace in the effort to reduce marine engine emissions. In 1998 the California Environmental Protection Agency's Air Resources Board (ARB or CARB) set standards more stringent than the federal ones to reduce hydrocarbon emissions by 75 percent in 2001 and 90 percent on 2008 models, five years ahead of targets for the rest of the country. "These new standards will deliver significant reductions in air and water pollution while still allowing Californians the full range of fishing, boating and other water sports experiences they now enjoy," said ARB chairwoman Barbara Riordan. As with federal regulations, there was no intention to require any modification or retrofitting of engines or watercraft sold prior to 2001. The cleaner engines will reduce smog-forming emissions by an estimated 110 tons per day by 2010 and by 161 tons per day by 2020. CARB analysts determined that switching from a two-stroke to a more efficient four-stroke 90-hp outboard engine would save the user more than $2,000 in fuel and oil costs over the average sixteen-year life of the engine, or $1,200 over a nine-year life. According to CARB, the new standards might "preserve water sport activities in areas where local water agencies have banned or are considering bans on boating activity because marine engines are polluting lakes and reservoirs."[54] Polaris Industries easily met the new standards before leaving the PWC market in 2004. In 1999, ARB chairman Alan Lloyd praised the company for meeting the standards with its Genesis PWC.[55] Still, the PWIA and outboard motor manufacturers sounded the alarm to forestall regulators from requiring cleaner, safer engines in the short term, claiming in one study that the standards meant economic doom for the industry because they would double the average retail price of an outboard motor to a staggering $14,000. No explanation of this doubling in cost accompanied the claim.[56]

Another environmental impact of PWCs is what beach- and lake-goers describe as their persistent, unpleasant, high-pitched noise. As with ATVs and snowmobiles, PWC noise has significant economic costs. Charles Komanoff and Howard Shaw point out that people don't like noise and will pay to avoid it; witness the reduced value of houses near airports and highways. In a study of PWCs published in 2000, Komanoff and Shaw first explained clearly the methodology they used to convert annoyance costs into dollar amounts, then they estimated two costs: the annoyance cost of jet ski noise and the expenses involved in strategies to reduce this cost. They concluded that the nation's more than one million jet skis impose approximately $900 million in noise costs on beach-goers each year, or an average of $47 per jet ski per day of use. With eighty thousand more jet skis manufactured and sold each year, the cost of jet ski noise could be expected to reach $1.07 billion by 2005, or 18 percent more than the total for 2000. These figures do not include the noise costs of reduced property values to residents of waterfront areas or the costs to canoeists, kayakers, and other boaters or to hikers on nearby trails.[57]

Jet ski noise, as Komanoff and Shaw point out, differs significantly from motorboat noise. Jet skis constantly bounce and skip across the surface of the water, which magnifies the sound in two ways. First, when the machine becomes airborne, the engine's exhaust, minus the muffling effect of the water, typically becomes 15 dB louder. Then when the machine hits the water on the way back down, it smacks it with an explosive "whomp," sometimes a series of them. Operators love to jump and bounce, running at full throttle to take advantage of the engine's power. They typically race around in circles jumping the wake, always in the same area, so that the noise never ends the way it does with a motorboat, which usually travels from one point to another and then anchors for fishing, sunning, or snorkeling. Moreover, rapidly varying noise such as that produced by a jet ski is much more annoying than constant unvarying noise. This suggests three strategies: develop quieter jet skis, require them to operate further from shore, or restrict their use to fewer bodies of water. Only the third approach holds real promise.[58] (Perhaps a fourth option exists: equip all beach-goers with headphones.)

The PWIA disputes these contentions. Its website indicates that both the National Association of State Boating Law Administrators (NASBLA) and the National Marine Manufacturers Association (NMMA) have enacted Model Noise Acts that the member manufacturers follow. These acts comply with noise standards set by the Society of American Engineers (SAE). NASBLA requires 88 dB for jet skis under the SAE J2005 standard and 75 dB under SAE J1970. NMMA recommends 90 dB under SAE J2005. Tests comparing noise levels emitted by 2001 models found that a three-seat PWC emits 70 dB at 100 feet when towed with the engine not running. When tested with a running engine at full throttle, the engine sound plus the water sound created 78 dB, well below the Coast Guard's boat noise regulation of 86 dB at 50 feet at full speed.[59] Even more, the noise of a PWC towed by a 150-meter rope without the engine running measured 68 dB at 25 meters at 44 kph, and 72–78 dB at 150 meters when running at full throttle.[60] The Environmental Protection Agency has determined that 75 dB at 50 feet is an acceptable noise level to protect public health and welfare, so currently manufactured PWCs come close to meeting these levels.[61] Yet the EPA lacks a noise abatement sector to establish meaningful standards and has never tested small-bore engines. In the absence of a federal body responsible for noise abatement, as a nation we have no means of determining objectively the costs and dangers of excessive noise. We rely on manufacturers to establish noise standards, and they are not required to share the results of their tests with consumers.

Unsafe at Any Speed, Revisited

PWCs were developed to meet postwar Americans' growing interest in machine-based recreation. PWCs have delighted enthusiasts with the thrills they provide and the access they give to beautiful, out-of-the-way sites. The quality and safety record of PWCs have not met appropriate standards, however. Roughly one of every five PWCs manufactured over the last decade has been recalled because of production or design problems—in most cases, problems that could lead to fires or explosions. According to the U.S. Coast Guard, between 1991 and 2000 nearly 300,000 of the 1.2 million PWCs sold were recalled. Bombardier recalled 224,113 units, Kawasaki 51,279, and Yamaha 10,725. The problems ranged from poorly designed fuel tanks to brittle and weak hoses as well as faulty O-rings and inadequate assembly methods. According to Coast Guard data, the number of both fires and injuries associated with PWCs increased 300 percent between 1995 and 1999. A total of ninety fires or explosions caused fifty-five injuries. In only six cases was the fire the result of operator error.[62]

Coast Guard data also confirm that during the years 2000 to 2003 more than 40,000 of roughly 250,000 PWCs produced (16%) were recalled because of production and design problems that could result in fire and explosion. Bombardier, Kawasaki, Polaris, and Yamaha have been most affected. Between 2000 and 2002 another sixty-six fires and explosions, the vast majority of them due to machine failure, caused another fifty-six injuries or deaths. Coast Guard officials have not pursued operator safety with adequate vigilance. For example, they recommended that a recall campaign involving 126,296 Sea-Doo watercraft be closed, even though nearly 80,000 had not been repaired. A closer examination of the "factory directed modifications" directed by the manufacturers indicates how serious the problems were. They included engine overheating sensors not monitoring the exhaust; the fuel tank vent hose malfunctioning and permitting fumes or gasoline to be expelled from the fuel tank into the engine compartment; and a series of other problems. Repair rates on recalls seem to be rather low generally.[63]

The response of the PWIA to published reports on these serious issues of quality and safety has been to proclaim that PWCs are "among the cleanest and quietest boats on the water today." The manufacturers, PWIA representatives contended, had been "unfairly assaulted when the anti-access group Bluewater Network issued a deceptive and misleading" report that cited voluntary recall figures as "evidence of production/design problems that could lead to fires." A PWIA press release issued on October 16, 2001, focused not on the recalls but on investments that would lead to product improvements. It noted that "glitches are inevitable" in a dynamic industry, and pointed out the prompt and responsible manner in which the recalls were instituted. The Bluewater Network ignored, PWIA insisted, that only 6 percent of vessel fires in 1999 involved

PWCs and the rest other marine vessels. Further, while 10 percent of all motorboats experienced fire-related problems in a year, PWCs had only eighteen fire-related problems annually. PWC use had increased significantly, yet 99.7 percent of them had never been involved in any accident of any kind.[64]

Most Americans believe that accidents are your own fault. Too many of us stand on ladders above the rung marked "Warning: Danger: Do Not Stand On or Above This Step." Altogether too many of us use lawn mowers, line trimmers, circular saws, and so on without safety glasses or ear protection or while wearing open-toed shoes. Millions of us still drive without using seatbelts, somehow believing that ejection from a vehicle during a crash is better than being trapped in a wreck should a fire ensue. So, too, PWC manufacturers argue that accidents are part of the risk of jet-skiing and that safety is the responsibility of the operator, as indicated in the vehicles' operating manuals and warning decals. Manufacturers contend that the operator's individual choices and behaviors are far more crucial to PWC safety than anything about the machine itself. And, like ATV manufacturers, they worry that needless government intervention will dumb the machine down, making it no fun whatsoever. Government intervention should be a last resort.

Although the number of recreational boating fatalities has been declining, the number of PWC fatalities has increased. Most of the deaths result not from drowning but from blunt trauma injuries. There are several factors that can cause accidents, including alcohol, overconfidence, and insufficient knowledge or equipment (no helmet or life vest). As with other recreational machines, most jet-skiers just hop on the vehicle and set off into the wild blue-green yonder. They don't take recommended safety courses. Safety courses are generally not required, though perhaps they should be. Or perhaps liability insurance should be required, with operators who take safety courses receiving a discount on premiums.

A close analysis of the PWC safety record indicates that there is an inherent level of danger involved in their use and that most accidents, injuries, and deaths occur during proper use by safe operators. PWCs can exceed 60 mph, but even at 40 mph a PWC travels about 60 feet per second. Couple this with the phenomenon of "off-throttle steering," and the situation can be dicey. According to the National Transportation Safety Board, the term *off-throttle steering* "is an oxymoron because there is little or no steering capability when the throttle is off." This counterintuitive situation no doubt contributes to some accidents. As the NTSB reports:

> When a new rider realizes there is danger of hitting another vessel or object, the operator's typical response based on experience with other motor vehicles is to first let off the throttle and then attempt to steer away from the hazard. But closing off the throttle leaves the vessel coasting in the original direction based on the effects of momentum, and without throttle there is very limited steering control. Personal watercraft have no braking mechanism; they coast to a stop and, while coasting, there is no turning ability.[65]

Steering and braking difficulties are rife in many of the accident reports of PWCs, and the PWC education programs that manufacturers endorse have done little to reduce the problem.

Unlike traditional boating, where falling overboard or swamping are considered accidents or even emergency situations, in PWCs they are expected events, for PWCs are designed and constructed to capsize on a regular basis, with most models having safety lanyards or "kill switches" that are connected to the operator's wrist to stop the motor when the operator falls off.[66] This is the only type of recreation vessel for which the operator is expected to fall off and for which the leading cause of death is not drowning but trauma. With trauma, the chance of survival is lower in a PWC accident than in a typical boating accident. PWC operators are more inclined than other boaters to wear personal flotation devices, but these may not help them in an accident, since they will suffer from contusions and lacerations to the head, face, and upper body when flung from the PWC into the water.

In 1996 there were 57 PWC-related fatalities, 1,831 injuries, and more than 4,000 PWCs involved in accidents.[67] Granted, the cause of many PWC accidents is inappropriate speed, inexperience, and inattention. As with automobiles, young persons are far more likely to be involved in accidents. More than half the PWC accidents reported in the first six months of 1997 involved operators between the ages of eighteen and twenty-seven; 143 of them were in the nineteen-to-twenty-one age group.[68] According to most estimates, over one-third of operators involved in fatal accidents were known or presumed to have consumed alcohol before the accident. With or without alcohol, with or without PFDs, too many operators drive improperly or carelessly. They ride near beaches and annoy beach-goers with noise as they weave, zigzag, and jump wakes, in so doing endangering themselves, other operators, other boaters, and especially swimmers.[69]

Educational programs in areas other than "off-throttle steering" might reduce accidents, but operators seem loath to take advantage of them. According to the NTSB, more than 80 percent of the recreational boat operators involved in fatal boating accidents had never taken any type of boating education course. Many of the accidents involved individuals who had rented the PWCs and embarked on joyrides after only cursory verbal instructions on how to operate the vehicles. The American Red Cross has reported that most PWC users have little or no experience. Nearly one-quarter of the PWC operators involved in accidents in 1997 were renters; 68 percent of the operators of rented PWCs were under twenty-five years old, and 73 percent had been on the water for less than one hour when the accident occurred. This would indicate that strict training and licensing programs might have some impact on accident rates. Indeed, at least twenty states have established rules to limit PWC rentals to persons aged sixteen or older and to require operators to undergo prior training, including video and other instruction, though not always on-water training. A number of states allow children as young as ten to operate PWCs.[70]

By 1998 there were about a million PWCs in operation, and they represented more than one-third of all new recreational boat sales. The number of recreational boating fatalities has been going down overall, but PWC fatalities are going up. Safety experts disagree as to whether this means that PWCs are inherently more dangerous than motorboats. A major reason for this disagreement is that data on accident and fatality rates are incomplete. For example, the Coast Guard cannot tell whether all operators comply with the law requiring them to report accidents in which death, personal injury, or property damage greater than $2,000 has occurred. Further, the Coast Guard normally gathers statistics only on numbered boats (categorized by length and hull material) operating in coastal waters, whereas many of the boats and PWCs involved in recreational accidents are not numbered, even if they are registered in a given state, and many of the accidents occur on inland bodies of water that do not fall within the scope of the Coast Guard's studies. As a result, its Recreational Boat Casualty Reporting System fails to include every accident involving a recreational vessel; the Coast Guard calculates underreporting at 6 percent.[71] Finally, experts disagree on what basis to compare boating accidents. Should one consider the length of the period of operation (hours per day or days per year), on the assumption that the longer the period of use of the vessel, the higher the chance of an accident? Are PWCs used for shorter periods than other boats? PWC enthusiasts argue that increases in injury and fatality rates have accompanied even higher increases in sales and use, so that in fact accident rates per numbers of PWCs, hours of use, and so on have declined.[72]

PWIA representatives argue that nonfatal PWC accidents are probably reported at higher rates than accidents involving other motorized watercraft. Among other things, PWCs are rented more than other boats, and PWC rental operators are likely to report accidents for insurance and product liability reasons. Also, many PWC accidents involve collisions, which most state laws require be reported. In contrast, people tend not to report it as a boating accident when they fall down in an open boat, hurt themselves while starting an outboard motor, or suffer an injury while canoeing or kayaking.

Finally, as with ATVs, increased numbers of vehicles may be the most important factor in the increased number of accidents. In 1987, some 93,000 PWCs were in use across the nation. The 376 PWC accidents reported in 1987 indicate an accident rate of 4.05 per thousand vehicles. Of those 376 accidents, 156 resulted in an injury and 5 were fatalities, for a national injury/death rate of 1.68/0.05 per thousand. In 1993, there were roughly 455,000 PWCs in use and 2,236 total accidents, for an accident rate of 4.91 per thousand. That year, 915 accidents resulted in injury and 35 were fatal, yielding a national injury/death rate of 2.01/0.08 per thousand. And, according to the PWIA, in 1999 there were approximately 1.1 million PWCs in the United States and 3,374 PWC accidents nationwide, resulting in an accident rate of only 3.07 per thousand. With 1,614 injuries and 66 fatalities that year, the national injury/fatality rate dropped to a low of 1.47/0.06 per thousand. To put it another way, the fatality rate per thousand PWCs has dropped, but the absolute number of fatalities and injuries continues to increase.

Even if fatality and accident data indicate "no significant statistical increase," the Coast Guard reported that in 1997 more than half of all serious recreational boating collisions involved one or more PWCs, even though they accounted for fewer than 10 percent of registered watercraft in that year. Hence, on the basis of serious collisions per hull, jet skis have an accident rate a dozen time higher than that of other boats. Even if they are "used" more actively than other boats, say twice as much, they would still have a collision propensity six times that of other boats. And as for boating fatalities in 1997, over 10 percent were to users of jet skis, which accounted for only 4 percent of recreational marine vehicles, and three-quarters of the deaths involved causes other than drowning (such as blunt trauma), a rate triple that for other boating deaths.[73] These data seem to hold on a state-by-state basis, too.

Industry representatives argue that hours of operation should be the standard for comparing PWC accident rates with rates for other motorized vessels.[74] They cite an NTSB study indicating a lack of data documenting comparable use. According to this study, "Riding time is an important factor in interpreting accident and injury information. To accurately compare PWC accidents to accidents involving other types of recreational boats, it is necessary to qualify the usage time by vessel type. If PWCs are used more often than other types of boats, then their exposure time for incurring an accident would be higher." In addition, conventional vessels spend much of their time docked, anchored, or drifting. They are "destination oriented," used to get from one point to another in trips of relatively short duration, while PWCs are almost always in active operation, jetting around at high speeds. PWCs are on the water as much as three times longer than most other types of boats.

But from the numbers of accidents and their seriousness, we may conclude that PWCs are more hazardous to operate than other pleasure craft, precisely because they were created for thrill activities that endanger even the most competent operator. The number of fatalities between 1990 and 2000 involving numbered boats has ranged from a high of 924 in 1991 to a low of 701 in 2000, a range of 8.3–5.5 fatalities per 100,000 numbered boats. In 2000, out of the 12.8 million numbered boats, 543,000 were PWCs, 93,000 of them in Florida, 41,000 in Ohio, 37,000 in Missouri, 34,000 in Minnesota, and 32,000 in Georgia, with the others in the remaining states.[75] There were 144,000 PWCs registered in Florida in 1999, so perhaps as many as 50,000 were not numbered. Given that the Coast Guard mostly gathers statistics on numbered boats, the numbers of PWCs involved in accidents, nationwide and by state, have almost certainly been undercounted. In Florida, in any event, PWCs account for roughly 11 percent of registered vessels but are involved in 24 percent of reported accidents.[76]

Front-line medical personnel have long worried about the epidemic of deaths and injuries accompanying the spread of recreational machines throughout the nation. Whatever the machine, and whatever the ecosystem, geography, geology, state, climate, or season, hundreds of thousands of operators have been seriously injured in accidents involving recreational machines. PWCs are associated with especially severe injuries.

In a 1993 study based on 1989 data gathered from eight midwestern states, surgeons at the University of Louisville School of Medicine determined that PWC accidents frequently resulted in severe injury, often to underage operators.[77] These injuries put a significant burden not only on families but on the medical personnel and resources of community hospitals.[78] Once the federal government began a systematic effort to gather PWC accident data, analysts recognized a growing crisis. Between 1990 and 1995 the number of PWCs grew threefold while the number of injuries grew more than fourfold. In 1995 the number of emergency room visits for injuries related to PWC use was 8.5 times higher than the number for injuries related to motorboats.[79]

PWC-related injuries are significantly different from the injuries typical of other boating accidents. One Florida study found that PWC accidents frequently result in brain injuries, liver, spleen, and kidney injuries, and skull and skeletal fractures.[80] Data from an Arkansas Game and Fish Commission study for 1994–97 indicated that personal flotation devices were effective in preventing deaths and morbidity. The Arkansas study reported 126 incidents involving 141 vessels and resulting in over $156,000 in property damage, with almost two-fifths of users injured, mainly with head trauma and fractures to the lower limbs. The researchers concluded that the trend toward collisions and injuries would continue, suggesting a need for changes in policy, education, and manufacturing standards to provide for safer operation.[81] Another study concurred that "the number of these injuries seen in hospital emergency rooms will most likely increase in the future as the popularity of water-related recreational activities becomes even more widespread."[82]

If a passenger survives an accident involving a PWC, he often faces serious physical pain from the extraordinary injuries he has sustained. Collisions and falls often result in head and neck trauma, facial fractures, lacerations, and concussions. One especially gruesome injury that can result from PWC accidents is "rectal blow-out," the effect of what is essentially a water enema administered at 70 mph.[83] Doctors have also noted that instances of "open-book pelvic fracture," a straddle injury that can result from a jet ski accident, seem to be on the increase.[84] Between 1996 and 2001, physicians at the R. Adams Cowley Shock Trauma Center in Baltimore, Maryland, treated more and more patients with direct collision injuries, handlebar straddle injuries, axial loading injuries, and hydrostatic jet injuries. Traumatic brain injury occurred in 54 percent of patients and spinal injury in 29 percent. The doctors noted that inexperience and reckless behavior were the principal factors contributing to the accidents.[85]

Safety laws and education help. Because of comprehensive PWC laws enacted recently in Florida, for example, while PWC registrations have increased by over 50 percent in the ten years since 1995, PWC accidents have declined over 67 percent in the same period. The PWIA attributes this record to laws requiring mandatory boater safety education for all boat operators, including PWC operators, under twenty-two years of age; mandatory PFD use; and mandatory use of an engine-cutoff switch lanyard (if available from the manufacturer). The state has also stipulated that weaving through congested vessel traffic, jumping wakes, and other dangerous behavior constitute "reckless operation of a vessel (a first degree misdemeanor)." Operators must be at least fourteen years old, while livery (rental) services must provide on-the-water demonstration, must evaluate the operator's comprehension of safety issues, and cannot rent to anyone under eighteen years of age. In Pennsylvania, a law introduced in 2000 mandating an eight-hour education course for anyone wishing to operate a PWC has led to a dramatic drop in accidents.[86]

Clubs and Ecosystems

Trade organizations that represent the PWC industry and local clubs work together to maintain access to recreation areas, make jet-skiing a family sport, and promote water safety. The major trade organization that seeks to maintain or expand access to bodies of water is the Personal Watercraft Industry Association, founded in 1987 as an affiliate of the National Marine Manufacturers Association. Manufacturers recognize their responsibility to produce PWCs that are safe to operate and to educate operators about safe operation, and they work through the PWIA toward this end. The average PWC owner is a weekend user, a member of the middle class, and usually a white American. According to a 1996 PWIA survey, PWC owners (not necessarily synonymous with PWC operators) on average were forty-one years of age and had an annual household income of $95,400. Nearly three-quarters of them were married, 40 percent were college graduates, 85 percent were male, 68 percent had owned powerboats before buying PWCs, and 73 percent used the vessel an average of 36.5 days per year.[87]

Jet ski clubs, like jet skis, are most widespread in Florida. Like their ATV and snowmobile counterparts, jet ski clubs promote training, safety, responsible riding, and environmentalism. They sponsor racing and slaloming events. Members pay dues, and the clubs get sponsorship funds from various local businesses. Club members have a series of agendas. One is to encourage family activities involving PWC weekends. Another is to combat what they consider to be unfair legislation that limits access to launches and sites. Most jet ski clubs also encourage safety by seeking to set voluntary speed and noise limits in congested areas, and by urging their members to become certified as safe operators. Many clubs require their members to seek certification. For example, the Coastal Carolina Jet Sports Club won a Grassroots Grant from the Boat/US Foundation for Boating Safety to post signs in the area around Charleston, South Carolina, advising PWC operators to boat responsibly.[88] The Space Coast Jet Riders in Florida describe themselves as a group of "fun-loving, responsible personal watercraft . . . riders, always looking for new waters to explore and new friends who 'support the sport.'" Their activities take them to the Indian and Banana rivers, Sebastian Inlet, Silver Glen Springs, St. Augustine, St. John's River, the Suwannee River, Jensen Beach, Lovers Key, Key Largo, and "even the Bahamas!" (All of these areas have environmentally important estuarine and riverine ecosystems, which raises the issue of what "responsible" riding entails.) Clubs seeking to foster a sense of environmental consciousness hold volunteer local cleanups, like those for highways and neighborhoods, although only a few dedicated members seem to show up. The Space Coast club features a photograph on its website of a manatee zone that requires boaters of all kinds to maintain slow speed and avoid wakes.[89]

The PWC code of ethics, as promoted by the PWIA, leaves no doubt what responsible clubbers have in mind:

1. I will respect the rights of all users of recreational waterways, both on public waters and adjacent private property.
2. I will be considerate of other users at the launch ramps and docks.
3. I will follow the navigation rules of the road around all other vessels, including regulations prohibiting wake jumping.
4. I will give all anchored or drifting vessels plenty of room.
5. I will always operate at headway speed in "no wake" zones.
6. When approaching shore, I will be especially aware of swimmers, divers and other craft.
7. I will not disturb wildlife and will avoid areas posted for the protection of wildlife.
8. I will not litter the shore, nor be careless with fuel or oil.
9. I will volunteer assistance in case of emergency.
10. I will determine my speed by my equipment, my ability, the weather, wave conditions and especially other vessel traffic.
11. I will not interfere with others' boating pleasure.
12. I will pay close attention to the noise level of my PWC and be aware of how others are reacting.[90]

However, the extent to which PWC operators follow such codes remains unclear. Public complaint about the "nuisance" of PWCs suggests that too few operators observe a code of ethics, and that too many lack training. To combat these problems, the PWIA has sought to develop safety and education programs. It works with state governments and other associations to establish more effective safety regulations and campaigns to change the image of PWC operators as reckless joyriders, at the same time lobbying manufacturers to improve the safety of PWC designs. This approach seems to have worked in states where it has been employed: in Pennsylvania, Minnesota, Wisconsin, Virginia, and California, accident rates have declined since mandatory programs were introduced.[91]

Like many ATV clubs, many PWC clubs exist principally to promote high-speed high jinks. For example, members of a New York and New Jersey PWC club explicitly describe how to complete such "potentially dangerous" tricks as the tailstand, barrel roll, bare-footing, reverse 180, turning sub, fountain, and hurricane. For the tailstand, "Pull back hard on the handlebars and give the boat some gas. . . . Now that you have the bow pointed to the sky, it's all up to you." For the hurricane, "Grab the throttle side of the handlebar with your right hand, and the rear boarding handle with your left. Jump off the right side of the ski and begin to turn as you give the craft gas. Try to keep your body as close to the handlebars as possible. Your [PWC] should be going in a very tight circle around you. Hold on tight and give the ski more gas to make it spin faster and faster."[92] The Jetty Jumpers of Brooklyn share the same desire to escape gravity:

> We are a club of Brooklyn extremists, testing the bounds of gravity and feeling Mother Nature at her best. All is still, the world at a pause. Able to be free to test your fears and self-discipline. Gravity pulls and you fight the tide. Your manhood is tested. Each wave has its own power and uniqueness, a style all its own. The more you experience the free-falling feeling, your life will alter. The Jetty Jumpers are starting to get their adrenaline higher and the eagerness for more air becomes immense. Sky's the limit and there is no stopping us. We will thrust and show style with our every move.[93]

Like ATVs and snowmobiles, PWCs bring people together into clubs with others who share similar interests. But whatever their weekend activities, the nature of their machines has drawn them into conflict with other outdoor recreationists. How can so many millions of high-speed vehicles be accommodated? Restrictions on their use have perhaps become inevitable, although every decision to restrict or regulate PWCs has generated ill feelings on all sides. Conflicts remain strong over access to natural resources. Citizens for Florida's Waterways published a list of groups that filed lawsuits to keep "family boaters, fishers and water-skiers of Florida waterways, supposedly in the name of manatee protection," and urged members not to contribute to these organizations, which were "systematically destroying family boating in Florida." The groups included the Florida Public Interest Research Group

(PIRG), the Sierra Club, the Audubon Society, the Humane Society, and the U.S. PIRG.[94]

Hullism: Attempts at Regulation and Legislation

On the basis of PWCs' alleged noise, polluting effects, and inherent lack of safety, in November 1997 the Bluewater Network sent a letter to the National Park Service requesting a ban of PWCs in national parks. In May 1998 the National Parks Conservation Association petitioned the National Park Service for an immediate ban on PWCs. In September 1998 the Park Service issued a proposed rule, which was followed by a comment period and a final rule in March 2000 that prohibited PWCs in Park Service areas unless the service determined that PWC use was "appropriate" for a specific area based on resources, values, other visitor uses, and overall management objectives. The rule provided two means to approve PWC use. The first, available to ten parks, was locally based and relatively streamlined, giving authority to the superintendent to make a ruling. The second was more formal and included such requirements as publication in the *Federal Register* and a public comment period. In the March 2000 rule, the service banned PWCs in sixty-six of eighty-seven parks, but gave twenty-one seashores, lakeshores, and recreation areas two years to establish regulations for PWC use. Any unit that wished to allow PWC use after this two-year grace period would have to undertake a complete environmental assessment. The grace period sparked litigation.[95]

Because of pressure from the PWC industry, the Park Service began to chip away at some of the restrictions on PWCs by reopening the process. On June 22, 2000, it announced a ninety-day public comment period on the agency's ban on PWC use on the Missouri National Recreation River. During this time, at the very least, PWC use would continue in the recreation area. The area, consisting of two sections of the Missouri River in South Dakota and Nebraska, was protected in order to "showcase a stellar example of a Great Plains river." The native flood plain forest, tall- and mixed-grass prairie, and the river itself provide habitat for several endangered and threatened bird and fish species, so PWC use in it might have significant environmental impact.[96] The decision to reopen the comment period occurred after PWC representatives claimed that South Dakotans and Nebraskans had had insufficient opportunity to be heard on the matter of restrictions concerning rivers in their own states. As Bluewater Network personnel pointed out, however, the Park Service had already held a ninety-day comment period, during which time the White House received over sixty thousand public comments, with more than forty thousand calling for a ban on the thrill craft throughout the system, including along the Missouri River. Were the Park Service to decide to allow PWC use on the Missouri, it would undermine the national scope of the decision and give impetus to overturn PWC bans in other magnificent parks.[97]

On August 31, 2000, the Bluewater Network sued the National Park Service and the Department of the Interior on two accounts. First, Bluewater asserted that the Park Service had violated the Organic Act and the Administrative Procedure Act when it issued the rule allowing PWC use in twenty-one areas for two years without banning them and without opportunity for notice-and-comment rule making. Second, Bluewater contended that according to the Organic Act, the Park Service had failed in its statutory mission to manage park areas so as "to conserve the scenery and natural and historic objects and the wild life . . . for the enjoyment of the same . . . as will leave them unimpaired for the enjoyment of future generations." The service's authorization of activities like PWC use was a "derogation of the values and purposes for which these areas have been established."[98]

Bluewater and the Park Service settled the case in April 2001, with the service agreeing to make changes in the final rule, primarily by requiring notice-and-comment rule making. Two pro-PWC groups, the PWIA and the American Watercraft Association, objected to the agreement, but their motion to have it enjoined was denied because the courts determined that the group did not have a legally protected interest in keeping parks open. Hence the decision to ban PWCs went ahead, with the majority of park superintendents banning PWC use over the next few months on the basis of their evaluations that PWCs endangered marine health, subjected operators and bystanders to risk of injury, and contributed to pollution.[99]

Industry groups hoped that their congressional allies would permit PWCs to continue to enjoy national park resources through legislation granting the Park Service a two-year extension of the grace period. In March 2002, House Republicans introduced such legislation to extend the grace period to the end of 2004 for those twenty-one parks still open to PWCs. The legislation did not pass. But on March 28, 2002, industry groups filed a motion in federal court to prevent enforcement of the final rule. In this motion the American Watercraft Association and the PWIA claimed that park officials had ignored their own procedures in banning PWCs from a number of parks without first completing required environmental assessments, and that they had acted arbitrarily and capriciously by regulating PWCs differently from other watercraft simply on the basis of hull type. The plaintiffs asked for a preliminary injunction to stay the bans on PWC use scheduled to take effect in thirteen national parks on April 22, 2002, and in eight more on September 15. The PWIA called the ban "an attack on our waterways" launched in "an eleventh hour, closed-door deal with the Clinton administration" by "an extremist, anti-access group dedicated to ending a wide array of recreational activities in the national park system."[100] PWCs, the trade groups pointed out, were "affordable family boats that seat up to four people and have no exposed propellers."[101] The lack of propellers meant that these vehicles "do not harm sensitive marine life."[102] To allow other motorized recreational boats but not PWCs was, apparently, pure hullism.

The trade groups contended that their suit was intended simply to ensure fair access to recreation sites. Monita Fontaine, then executive director of PWIA, said, "It is fundamentally unfair to arbitrarily exclude people from enjoying these public waterways without due process. Complete the studies, then decide."[103] Fontaine explained, "We're not saying that personal watercraft should be allowed in every park. . . . Clearly, each park is unique, and motorboats may not be appropriate in some environments. But we are confident that objective, scientific studies will find that today's personal watercraft have come a long way from those sold just five years ago and are among the most environmentally-friendly motorboats on the water. We welcome the National Park Service's scrutiny."[104] Naturally the trade groups welcomed such additional scrutiny, for the delay involved in preparing the requisite studies could only work in their favor, the Bush administration having already shown its willingness to ignore, reject, and weaken existing environmental protection laws.

A federal court denied the injunction, for several reasons. The court said that despite improvements in new PWCs pertaining to speed, noise, maneuverability, and pollution, the industry groups gave no evidence of the proportion of new to older PWCs in use, and hence could not claim broad-based, satisfactory improvements. The court also ruled that the Park Service could attack the problem in this way since this action involved offering considered responses to comments during the public comment period. The court finally held that PWC users are responsible for choosing where they operate their PWCs. They could still use park areas, just not on their PWCs, and there was no basis for a claim of loss.[105]

In April 2002 the Park Service announced permanent PWC bans for five of the remaining twenty-one parks: Cape Cod and Cumberland Island National Seashores, Delaware Water Gap and Whiskeytown-Shasta-Trinity National Recreation Areas, and Indian Dunes National Seashore. PWCs were banned at the other sixteen units pending further study later in 2002.[106] Unfortunately, in 2003 the Department of the Interior issued a mandate requiring that NEPA (National Environmental Protection Act) analysis be undertaken in individual areas to determine whether PWC use should be allowed by special regulation, despite earlier determination that PWC use was inappropriate. This mandate, according to one observer, "represents a major shift in the balance between conservation values and recreation values. I have been unable to locate any documents explaining the change in policy."[107] The mandate may have been unprecedented. This observer continued, "NEPA analysis is generally undertaken when a proposed government action will have an adverse effect on the environment. In keeping with a commonsense reading of the statute and its purposes, government actions to protect the environment do not require NEPA analysis. Yet that was what [Interior mandated] in this situation." This was "a highly creative interpretation of NEPA, to put it mildly."[108]

Local and state officials trying to control the impact of PWCs on people, wildlife, and places are hamstrung by the confusion at the federal level as well as by congressional efforts to protect the interests of recreational machine users (and manufacturers). When elected officials or regulators seek to promote environmental protection and recreational machine safety, manufacturers and distributors understandably seek to shape the rules so as to maintain their profits. In this effort they generally raise three arguments. First, they protest that poor legislation will bring about economic doom; they proclaim themselves unable to meet new standards even while their sales brochures promote their machines as being the most advanced in the world. Second, they dispute evidence of the threats to ecosystems that their machines present. Third, they argue that their machines are safe to humans if operated according to instructions.

Yet many local, state, and national groups dispute these claims, and in the absence of consistent and timely federal regulation and leadership, local communities have expanded their efforts to regulate PWCs. In Southern Shores, North Carolina, the town council took advantage of the power specifically granted to seventeen North Carolina oceanfront municipalities, including Southern Shores, to regulate jet ski operations. They took the action in 1994 soon after a fourteen-year-old boy was killed when the jet ski he was operating slammed into a bulkhead. Residents requested the enactment of an ordinance prohibiting jet ski operations within the town's interior canal system; prohibiting jet ski operations by any person under the age of sixteen; and prohibiting jet ski operations within 300 yards of any sound-side shoreline within the town and its extraterritorial jurisdiction, except for purposes of access and egress and then at speeds not to exceed 5 mph. In March 1996, after long deliberations, the town council enacted a fairly comprehensive ordinance addressing such considerations as personal flotation devices, lanyard requirements, permissible hours of operation, proscription of muffler modifications, minimum distances from other vessels, and wake jumping.[109] Townsfolk thereupon pushed for the jet ski ordinance be amended to establish several prohibition zones and to limit speeds beyond a 400-yard restricted zone but within the town's jurisdiction to 25 mph. In May 2004 most of these recommendations were accepted and passed into law.[110]

Other communities have banned or restricted PWC use as well. When challenged by industry representatives, the Washington State Supreme Court confirmed the right of county commissioners to pass a law banning PWC operation in and around the San Juan Islands. Lake Tahoe, California, also banned PWCs that did not meet EPA 2006 or CARB 2001 noise and pollution standards, a ban unsuccessfully challenged in the courts by manufacturers. Maine banned the use of PWCs wholly or partially on 245 "gem" lakes, and Vermont has banned them on lakes smaller than 300 acres.[111] In 2002, after a difficult and time-consuming process involving the state, the National Parks administration, and NGOs, four Cape Cod towns—Eastham, Orleans, Chatham, and Harwich—gained authority to ban PWCs.[112]

As noted earlier in the cases of Florida and Pennsylvania, since the 1990s several states have also introduced laws to regulate PWCs, restrict their access, prohibit underage operation, and require operator education programs. Not all of the legislation made it out of committee, and few bills were passed. The bills addressed safety and nuisance issues more than environmental concerns and habitat degradation. In 1999 the Arkansas General Assembly enacted a fairly comprehensive bill covering the safe operation of PWCs that mandated the use of PFDs and (if available from the manufacturer) lanyard-type engine cutoffs and restricted PWC operation to individuals at least fifteen years old and PWC rentals to individuals at least eighteen years old. The bill stipulated that operators must use the machine "in a reasonable and prudent manner." Nebraska followed suit, except for requiring fifteen-year-olds to attend a boating safety class and limiting rentals to those sixteen and older. During 1999 at least four state legislatures passed legislation that required persons aboard PWCs to wear PFDs. While at least six state legislatures considered a boating safety certification requirement during the 1999 legislative session, only one state, West Virginia, enacted a bill on the subject. At least seven states enacted legislation dealing with boating under the influence of alcohol or drugs, and at least three focused on the issue of vessel homicide.[113]

New York governor George E. Pataki signed into law legislation that allows local governments to regulate the use of personal watercraft up to 1,500 feet from shore on New York's waterways. The legislation allowed cities, towns, and villages to regulate or even prohibit PWCs in municipal waters following the holding of a public hearing and the adoption of a local law. According to Pataki's script, "This new law puts the power to decide what is best for a local community right where it belongs: in the hands of the people of the community." There were about fifty thousand PWCs registered in New York at the time the law passed, accounting for less than 10 percent of the recreational vessels registered in the state, but citizens were sufficiently outraged by them to support the legislation. Not surprisingly, clubbers and the industry saw the New York law as a bad precedent.[114]

Losing several battles at the local and state level, industry organizations turned to a decidedly more sympathetic Congress and president. On May 4, 2004, Brian Berry of the American Watercraft Association reported on a hearing in the U.S. House of Representatives calling the National Park Service to task for governing public lands according to "bias," not science. According to testimony of PWC manufacturers, operators, and club members, the Park Service had engaged in a "discriminatory ban" of PWCs from national parks. Representative Devin Nunes (R-CA), who called the hearing, chided the service for long overdue rule making connected with a decision to ban PWCs from parks, a process that had already taken two years. Nunes said, "This is simply a matter of fairness for American families. It is vitally important that our national parks be open and accessible to everyone, including those who want to use personal watercraft." While seven national park units had completed the rule-making process and permitted PWC users to come back, nine had not yet made a decision, and Nunes argued that users deserved to know "if they can enjoy the upcoming summer months on the water."[115]

The deputy assistant secretary of the interior, Paul Hoffman, who testified on behalf of the machines, reported that four units were in the final stages of rule making and might open any day to PWC use: Pictured Rocks Islands National Lakeshore (Michigan), Fire Island National Seashore (New York), Bighorn Canyon National Recreation Area (Texas), and Gulf Islands National Seashore (Florida and Mississippi). He also brought up the case of Biscayne National Park in Miami, Florida, where park managers had banned PWCs. Hoffman and the members of the subcommittee agreed that the ban should be reconsidered, and they pushed the park managers to reconsider their December 2004 rejection of a petition asking for a local environmental assessment of PWC use. Dave Bamdas, owner of Riva Motorsports in Pompano Beach and Key Largo, testified that "there was never any study at Biscayne, scientific or otherwise."[116]

PWCs on the Cusp of the Wave

A ban on personal watercraft has been in place since 1992 in the Key West National Wildlife Refuge and the Great White Heron National Wildlife Refuge. Great white herons, a white color-phase of great blue herons, are found only in the Florida Keys. They were driven nearly to extinction by the demand for feathered hats. These herons have successfully returned and enjoy feeding at dawn and dusk on tidal flats around hundreds of backcountry islands. Endangered green sea turtles and threatened loggerhead sea turtles successfully nest in the refuge, and hawksbill sea turtles feed in its beds of seagrass. Yet the PWIA challenged the ban on PWCs in this area on the grounds that it violates the state law forbidding discrimination against PWCs. Officials in the Florida Department of Environmental Protection seem unlikely to reverse the ban owing to overwhelming public support for it.[117]

Yet PWCs will find a place to play. From promising beginnings in the production of a new kind of motorboat, based on commonly available engines and pumps, the PWC industry has grown into a major player among recreational machine manufacturers. The companies have designed and produced faster, more exciting, and more powerful machines whose operators quickly grasped their essence. Like snowmobiles and ATVs, they epitomize the practices of Fordism and assembly-line mass production. The fiberglass-hulled vehicles, powered now by three-cylinder four-stroke engines, have changed the nature of marine recreation. This is a machine for slaloming, jumping, and creating wakes. It enables visits into habitats with shallow water. But the millions of people who have purchased these machines did so not to commune with nature but to speed through it. The machines are loud and intrusive and so have been banned in many park systems.

In response to persistent criticism, PWC manufacturers have redesigned their machines and claim that they are "one of the most environmentally friendly motorized vessels on the water." They dispute claims about "alleged harmful environmental impact, despite evidence to the contrary." They point out that each of the fifteen national parks to have completed an environmental impact study has determined that "PWCs present no significant unique environmental impact compared to other boats." Therefore, they argue, PWCs should no longer be banned from those parks if other boats are allowed. They note that most of today's PWCs use four-stroke direct-injection and two-stroke catalyst technology. They assert that many of the unburned gasoline and gasoline additives from two-strokes evaporate from water within the first hour after release. Moreover, today's machines are 70 percent quieter than those produced as recently as the 1990s because of hull insulation, exhaust system improvements, and new noise-absorbing materials. Many of today's PWCs do not leave the water at all because they are longer, wider, and heavier than earlier models, while new steering technologies "assist the operator in turning the vessel by continuing to supply thrust or activating small fins while the watercraft is decelerating."[118] But while one new PWC may not contribute significantly to noise, pollution, or ecosystem degradation, 1.5 million of them are another matter, and they may face a difficult future. As hard as their defenders try to promote them as family-oriented pleasure craft, to their detractors and to the courts, they are "thrill craft."

Ron Webber, a Maine recreationist, enjoys kayaking in Maine's isolated northern lakes and sitting quietly with his binoculars a few hundred yards from loon nests. He seeks the loons out but keeps his distance. One morning two jet-skiers interrupted Ron's tranquil vigil by buzzing in and out of the shallow end of a lake. Realizing that they were deliberately attacking loon nests, Ron reported them to the state wardens. The wardens investigated but said they doubted they'd be able to identify the culprits, who had disappeared at high speed. This menace to loons exists on all northern lakes, and the number of incidents is growing.[119] Until wardens have the resources to enforce laws—or better still, until jet skis have been banned from areas where machines and nature cannot coexist—the Ron Webbers of the world will continue to see machines and nature collide, with nature losing out.

◀ 5 ▶

SMALL-BORE ENGINES AROUND THE HOME AND GARDEN

CHECK YOUR GARAGE or your toolshed. If you are like most Americans, you have several small-bore engines lurking inside. Most of them will be dirty and dusty, out of tune, with spark plugs several years old, ensuring that they pollute heavily. They pollute heavily in any event because for the most part they still use two-stroke engines, which are cheaper to build and lighter than the cleaner four-stroke engines. They power lawn mowers, weed wackers (aka line trimmers), snowblowers, cultivators, chain saws, power brooms, edgers, hedge trimmers, and the hated and ubiquitous leaf blowers. They come in several varieties and sizes. Lawn mowers, for example, are ride-on and push, self-propelled and motorized, mulching and thatching, with bag and without. Ride-ons come with a variety of attachments so that they can double and triple as plows, graders, and cultivators. If manufacturers can motorize it, they will. And if it's motorized, Americans will buy it. Some people think the obesity crisis in the United States has much to do with how sedentary Americans have become. Perhaps the small-bore machines in their garages are a major source of their lazy behavior, since they no longer truly work around the home and garden as much as ride and wack.

Gardening machines offer great benefits in saving time and easing difficult tasks, yet like ATVs, snowmobiles, and jet skis, they also impose social and environmental costs. They create loud noise. They treat soils and the flora growing on them as inert substances to be rearranged, atomized, and pulverized as quickly as possible. They emit pollutants. Gardening machines represent the further industrialization, transformation, and degradation of nature. They have brought noise and dust into what used to be quiet residential neighborhoods. No longer can people recline in the hammock to enjoy the fresh air and the sounds of birds. In my neighborhood, people use power machines in their yards with a vengeance, from 7 A.M. until after sunset. The machines have one purpose: to attack what once was seen as organic material good for mulch or compost, blowing it away or collecting it to be placed in plastic bags for a trip to the landfill. A friend of mine suggested an appropriate hell for the users of leaf blowers, to me one of the most problematic of these machines: they would face an endless field of leaves, with two leaves replacing each one blown aside, and be required to blow them away using their own lungpower, puffing their cheeks out incessantly and for eternity. I try to avoid that purgatory by using a rake or just letting the leaves decay where they fall.

Whence and Whither the Lawn?

Americans have cultivated 40 million acres of lawn. This is more land under cultivation than for any single crop, including wheat, corn, or tobacco. Americans spend $1 billion annually on grass seed alone and roughly $40 billion on lawn and garden care pesticides, herbicides, and fertilizers. These highly toxic chemicals—after all, many of them are poisons—are essential accompaniments to lawn mowers and other gardening machines. The lawn, a monoculture preeminent, stretches from coast to coast and from north to south, even in such arid climes as Arizona and Las Vegas, where irrational and profligate use of water enables lawns to be established where they oughtn't otherwise grow. Lawns thereby homogenize the environment, from house to house, neighborhood to neighborhood, across the land, with the assistance of potentially dangerous chemicals and expensive machines powered by small-bore engines.

The lawn has always been, and is increasingly, a product of technology, a monoculture that requires constant watering, molding, shaping, cutting, de-thatching, fertilizing, and weeding. People see dandelions and crabgrass as dangerous invaders that must be eliminated. Cutting the lawn is something we all must do on a regular basis or risk the ire of neighbors. To the consternation of my neighbors, I have turned more and more of my yard over to what I take to be natural processes. I let grow what grows, and I rarely cut it. I like the complexity, whereas my neighbors do not. Most neighbors obsessively strive to create the best lawn using hybrid seeds and chemicals and machines, for the grass should always be greener in your own yard. In my case, the neighbors' grass always *is* greener. They look at my scraggly lawn, and the saplings and bushes that have begun to fill it, with a sense of puzzlement, concern, and misgiving. While the lawn on your property is clearly yours, your right *not* to cut it is less certain.

The lawn came to the New World with European settlers who brought seeds imported for agricultural purposes, and its evolution here reflected the rise of the lawn in Europe. By the first decades of the nineteenth century the lawn had already become an extensive phenomenon, but it did not become the focus of intensive study or construction until the rise of urban parks, especially as championed by such architects as Frederick Law Olmsted. Builders of lawns intended the creation of a pastoral paradise. In parks the lawn would draw Americans together; at home it would indicate prosperity. The lawns that typically surround government, religious, and cultural buildings reflect the power of those institutions. Which major church, town hall, or corporate center does not have a huge lawn that requires gallons of water, scores of pounds of chemicals, piles of dollars, and teams of machine-equipped gardeners to maintain them? How do city and state governments whose leaders claim budget deficits find the resources nonetheless to maintain lawns, even alongside highways and parkways that must be cut regularly? What is the Washington Mall but a green ribbon signifying the unified power of Congress and the White House? Occasionally too it welcomes protesters of various political stripes, but only if they can get permits.

The American lawn is a product of technology, horticulture,

genetic science, and applied botany, but above all, it is an industry. The journals of the industry—*American Lawn Applicator, The Greenmaster, Outdoor Power Equipment, Turf News,* and others—cover everything from pest management and grass seed hybrids to golf course design and sports turf. Indeed, the sports industry has been a driving force behind the multi-billion-dollar business of turf grass research and development. Land grant universities have received hundreds of millions of federal dollars to develop sports lawns. The Golf Association of America has also helped subsidize the development of the sports lawn. Researchers have studied grasses for the bounce and speed they confer, for their "torque," traction, and skidding properties, for their growing behavior and water requirements, and for their durability and color (to ensure lustrous greens on color TVs). They have developed strains of grass that can be cut low and still withstand weed killers, even if toads, snakes, nematodes, and other creatures cannot. Lawns can be grown from seed or built out of sod, at great cost; sod costs about $0.50 a square foot.[1]

Gardening machines developed in step with the rise of the American lawn as a cultural icon in the second half of the nineteenth century. They also developed in step with automobiles and the first recreational machines. Early lawns served municipal purposes, in the form of public parks, or symbolized the social status of wealthy citizens. The first lawn mowers were common laborers armed with scythes and such. After the Industrial Revolution, the task was increasingly turned over to mechanical devices. These machines were powered by men or animals. In the 1870s, inventors produced a two-man lawn-cutting device, a machine based on a cutting-reel blade and roller with one man pushing and the other pulling. One of the first gas-powered devices was the Multiplex Lawn Mower, capable of handling up to five cutting-reel units that increased the width of the cut to 12 feet. This turn-of-the-century machine "operated precisely like an automobile, with foot brakes and gear shifts . . . traveling from three to seven miles an hour." An early trimmer-edger existed as well. The Capital Trimmer had applications for edging drives and paths. The Pennsylvania Trio, a three-unit reel mower for large lawns, golf courses, and parks, cut a swath nearly three times as wide as a single unit, but it was horse drawn. The Triplex Lawn Cleaner used rotary brushes to sweep up the clippings. Coldwell's Combination Motor Lawn Mower and Roller had a four-cycle gas motor and a governor that maintained a constant speed.[2]

To keep park maintenance costs down, turn-of-the-century Chicago officials used sheep as an experiment in the 500-acre Washington Park. They found the results so impressive that they acquired more sheep for the next season. The sheep had the bonus of attracting tourists but the disadvantage of requiring a "flock-master" to keep them from wandering into the boulevards and to protect them from dogs. For applications after May 10 or so, they cleaned more neatly than a lawn mower could and could be sold for meat at a profit at the end of the season.[3]

The railway industry introduced a gasoline-operated railway weed mower in 1914 that eliminated "the expensive and slow method of cutting weeds along the railroad tracks with scythes." The machine—a motorized car equipped with two cutter bars like those of a farm mower and powered by a 6-hp engine—cut a swath 6 feet wide on both sides of the track. The car moved at 3 mph. A simple mechanism raised and lowered the bars to avoid obstacles. The inventors claimed that the machine would cut 25 miles of roadbed and right-of-way daily at a cost of $7 for fuel, oil, and the three men needed to operate the car, replacing one hundred men with scythes.[4]

Early promoters of the lawn mower believed that any labor-saving device was a good device. They saw only the aesthetics of green carpets of lawn and were unaware of the dangers—the decline of biodiversity and of sustainability—inherent in creating a monoculture of grass. Wrote one observer, "The well kept lawn is the only satisfactory carpet. A good stretch of grass is pleasing both to the eye and to the foot." To achieve that end required the heavy labor of cutting and rolling, but that task became increasingly motorized and, by the mid-1920s, a practical and manageable task within the budgets of most middle-class Americans.[5] Three quarters of a century later, the task had grown to huge proportions. By 2002, eighty-five million (or eight out of ten) American households gardened. Between 1997 and 2002, spending on professional garden and landscape construction grew threefold, from $3.6 billion to $11.2 billion. The average household spent nearly $500 annually on lawns and gardens, for a total of $39.6 billion in 2002 alone.[6]

Because of the simplicity of lawn and gardening machines, a large number of companies entered the business in the 1920s and 1930s, and many of them remain; there has been less shake-out in the gardening and snowblower industry than in the recreational machine industry, where only a handful of manufacturers exist for each kind of machine. Capital costs for tooling and retooling are substantially less than for recreational machines with their brakes, suspensions, transmissions, and such. You can easily create a tool and mount a small-bore engine on it to grind, churn, spin, cut, wack, blow, vibrate, clip, and chew in a variety of ways, with no steering, lighting, or suspension required (except on ride-on mowers). Several of the big names—John Deere, Sears, and others—were also involved in snowmobiles or ATVs for a while. Yamaha and Honda still are. The others—Stihl, Bobcat, Briggs and Stratton, Jacobsen and Lawnboy, Husqvarna, Black and Decker, and Homelite—focus on gardening, lawn, and lumbering equipment, although a number of them also build earthmoving equipment.

One such firm, Stihl, dates to 1925, when the German inventor Andreas Stihl, disturbed by what he thought was the primitive nature of logging tools and methods, designed and assembled an electrically powered chain saw that weighed 140 pounds. Loggers "loved it" because of its power and portability. He developed a gas-powered saw in 1929 and added automatic chain lubrication and a centrifugal clutch in 1934. In 1937 he visited the United States and Canada, immediately gaining a market in North America. In 1950 the company introduced its first one-man saw. Over the next three decades Stihl added a series of other improvements. In 1977 the company broke ground for a plant in Virginia Beach, Virginia, and now manufactures more than 1.5 million power heads annually for different models of electric and gasoline-powered tools.[7]

In 1952, representatives of the gardening equipment industry met in Washington, D.C., to establish the Lawn Mower Institute. In 1960 they changed the name to the Outdoor Power Equipment Institute (OPEI). Officers of such companies as Yard Man, Jacobsen, Eclipse Lawn Mower, and Toro determined to pool their resources to represent their interests before the government and the public in the promotion of walk-behinds, riders, tillers, snow throwers, shredders, leaf blowers, chain saws, and trimmers. The eleven charter members soon became seventy-seven, representing 98 percent of the manufacturers of outdoor power equipment.

In the 1960s OPEI focused on safety standards for power lawn mowers. The first standard required manufacturers to include an owner's manual in the box with each lawn mower. In 1964, in cooperation with the U.S. Department of Health, Education, and Welfare, OPEI produced its first safety video, "Mowing Lesson for Charlie." In the 1970s, as part of national trends toward consumer education, OPEI began to work with government agencies, including the newly created Consumer Product Safety Commission, on product liability, environmental, and safety issues. In the 1990s OPEI turned to emissions issues as pressure built from the Environmental Protection Agency to reduce power equipment pollution. OPEI established a Clean Air Act Committee to work with the EPA.

Another trade organization, the Portable Power Equipment Manufacturers Association (PPEMA), operated for forty-five years, originally as the Power Saw Manufacturers Association. The name was changed in 1982 to reflect the organization's growing scope. PPEMA voted to cease operations as of December 31, 2001, because of the overlap of its mission with that of the Outdoor Power Equipment Institute.[8]

According to the U.S. Census Bureau, there are nearly ten thousand farm and garden machinery and equipment wholesalers in the United States. About twelve hundred retailers sell lawn, garden, and farm equipment and supplies, plants and shrubs, cut flowers, fertilizers, and so on. There are approximately four hundred manufacturers of lawn and garden equipment, considering attachments, sprayers, tillers, chippers, blades, and bags, not just mowers, trimmers, cutters, chain saws, blowers, slicers, and dicers. These machines are quintessential American technology: time- and labor-saving, relatively inexpensive to operate, and easy to mass-produce. The rake, broom, and hose, although more sound from an environmental point of view, have been largely relegated to the compost heap of history. Given the large number of manufacturers and retailers, the number of machines sold annually, and the number of Americans who own them, there can be no surprise over the noise, pollution, and public health concerns that many citizens now have.

A significant problem associated with the industrial lawn is its high environmental costs. The inputs to an industrial law include fossil fuel energy, irrigation water, fertilizers, pesticides, and the grass seed or sod itself. Lawns have a net carbon loss when the carbon in the fossil fuels used in lawn care is counted. Lawns have such a structure that they generate more surface runoff than other home gardens, which means that more nutrients are lost in drainage water, and pesticides and fertilizer nutrients also wash into neighboring water supplies. To make things potentially worse, most lawn care pesticides remain untested for their long-term effects on humans and the environment, and roughly half of Americans do not read the warnings on containers. Commonly used lawn care chemicals have often been implicated in health problems. And over forty years after the publication of Rachel Carson's *Silent Spring,* chemical manufacturers still "tend to dismiss reports of illnesses or death as isolated and unsubstantiated cases." The ultimate results of industrial lawns are less biological diversity, increased global warming, stresses on municipal water supplies, increases in solid waste problems, pesticide contamination of food chains—to put it simply, pollution and public health problems.[9]

An alternative for some people would be to use goats, and several firms seriously offer the service. Though neighbors might object to the presence of livestock in their communities, goats can be organized into herds of various sizes for use by homeowners, private land managers, and public agencies to graze sites ranging from neighborhood yards to 30,000-acre ranches. The cost is roughly $700 per acre per year, and according to one goat-grazing firm this includes the cost of transportation, the shepherd's salary, supplements and health care for the goats, fencing, and insurance. "The goats will eat most vegetation that is available, including plants such as poison oak that are difficult to clear by hand. They will readily consume otherwise undesirable species such as pampas grass, any kind of thistle, and blackberry. By generally eating the top of the plant instead of removing it by the roots, goats may be less damaging to native plants when compared to traditional grazers." Just perfect for the lawn.[10]

Ode to a Leaf Blower

Most Americans own not only a motorized lawn mower but an assortment of gasoline-powered devices including trimmers, pulverizers, wackers, and above all leaf blowers. The first leaf blowers appeared during the nineteenth century, although these versions were innocuous: hand-held bellows used by Japanese gardeners to remove leaves and twigs from moss-covered soil. Japanese engineers attached hoses and blowers to powerful electric motors around 1970, with gasoline-powered blowers introduced soon thereafter in the United States. Droughts and other weather conditions in California facilitated the initial acceptance of the leaf blower: gardeners were prohibited from using hoses to wash away garden debris. Blowing the debris away solved many problems, although it created others. By 1990, annual U.S. sales surpassed eight hundred thousand units.

The weed wacker or line trimmer also appeared in many garages in the 1970s. The first weed wackers were heavy, with some gasoline-powered models exceeding 30 pounds. Engines ranged in size from 14 cc to 30 cc, and the machines sold for as little as $35 and up to $300 for the biggest ones. Electric-cord and rechargeable models followed. One innovation was placing the motor at the top rather than at the base; this made the unit well balanced and easy to hold. An automatic head for feeding the nylon was another innovation, as was the "Tap 'n Go" head. The Weed Eater company alone has a dozen different models. By 1980, twenty manufacturers were producing line trimmers, and there were already about twelve million in use.[11]

Articles describing early models were as a rule accompanied by photographs showing gardeners using the equipment without gloves, safety glasses, or ear protection. Safety was not a major concern. The selling point was ease and efficiency. As the editor of *Farm and Power Equipment* noted, when you use a trimmer "you're in for a pleasant surprise. Line trimmers cut grass, weeds, and other vegetation." The units handled tight spaces "along walls, around rocks, trees and posts, under fences and shrubs" well. The electric models were quiet, nonpolluting, and extremely safe, he wrote.[12]

George Ballas, a Houston inventor, patented the first nylon-line weed trimmer in the early 1970s. Ballas had thought of this new way to trim grass and weeds while sitting in his car at a car wash. He noticed that the whirling nylon bristles that spun along the top of his car cleaned it without causing any damage to the vehicle. He reasoned that a nylon cord might trim brush and weeds around trees without damaging the bark. He took an empty popcorn can, punched holes in it, put fishing line through the holes, and replaced the metal blade on his trimmer with the can. In 1977 he sold his "Weed Eater" to Emerson Electric of St. Louis, which marketed it successfully. People liked weed wackers because they could be used relatively safely by all members of the family. In the late 1970s, Black and Decker, Toro, and Sears entered the market, and advertisements began to appear on TV. Weed eaters replaced hand clippers and cordless electric grass shears.[13]

Leaf blowers epitomize false efficiency. Manufacturers developed them to sweep up such debris as leaves, twigs, grass clippings, and dirt in greater quantities and at greater speeds than a rake or broom might achieve. But leaf blowers in fact permit operators to perform clean-up tasks at rates only slightly faster than humans working with ordinary rakes and brooms. The machines are inorganic in every sense of the word. They destroy or weaken the plants whose ornamental value in the garden or park they are intended to enhance. They blow at speeds of up to 180 mph and can rip leaves from branches, critically injuring new growth, and disperse humus far and wide. Blower wind causes dehydration and leaf burn, and it suspends photosynthesis and other natural plant functions. According to Steve Zien, executive director of Biological Urban Gardening Services and owner of the Living Resources Company, this ruins the ecosystem: "Overall growth is slowed . . . disease spores laying dormant on the soil or fallen debris are blown back onto the plants . . . blowers effectively distribute disease spores, weed seeds and insect eggs throughout the landscape."[14]

One of the reasons for the widespread acceptance of leaf blowers and weed wackers is that many professional landscaping companies see the leaf blower as the quintessential American technology: time- and labor-saving, relatively inexpensive to operate, and simpler to deal with than human workers. American businesses, government offices, and others have all joined the wacking band. Municipal parks-and-recreation crews, professional gardening crews, universities and colleges, businesses, and individual households gather the detritus they have blown into a pile, throw it into the back of a truck, and cart it to a landfill where it does little as it decays to help the environment. Or they place it in plastic bags so that the organic matter serves no biological purpose. This organic material would be better used as mulch or turned into fertilizer. Instead, it has become a large component of the weight and mass filling transfer stations. In its place, crews, businesses, and families must use chemical fertilizers to feed plants denied the organic material of mulch. Without mulch, erosion, evaporation, and disease all become greater problems, and landowners must purchase commercially produced mulch—cedar chips, bark, peat moss—to make up the difference.

As with recreational machines, manufacturers exaggerate, underreport, or otherwise mislead consumers about the noise levels of gardening machines. Manufacturers report leaf blower noise at 50 feet in the range of 70–75 dB. But leaf blowers are routinely used at less than 50 feet "from unconsenting pedestrians and neighboring homes that may be occupied by home workers, retirees, day sleepers, children, the ill or disabled and pets." And they are used right next to the ear of the user.[15] Further, the sound of a leaf blower at 65 dB in no way resembles conversation at 65 dB, and you would not want that noise in your home. According to *Consumer Reports,* no backpack blower on the market meets this standard. The backpack models are even closer to the operator's ears and heart—mere inches away. According to Dr. Robert Blum, "Vibration is significant because commercial blowers are worn on the back. . . . Vibration is transmitted up the spinal column to the skull and temporal bones, which enclose the cochlea. . . . Ear muffs do nothing to protect the operators from vibration transmitted by bone conduction. . . . Vibration-induced hearing loss is well-documented."[16]

The typical noise level for gas-powered lawn mowers is 85–90 dB; the level increases to 95 dB for riding mowers. The World Health Organization and the EPA recommend that people limit their total exposure to mowers to forty-five minutes a day for the quietest ones, fifteen minutes a day for the average ones, and five minutes a day for the loudest ones. Few U.S. homeowners and few if any landscaping companies so limit their use.

Noisy gardening machines put U.S. manufacturers at a disadvantage on world markets. Quieter lawn mowers have been produced in Europe for years. Were U.S. manufacturers—either voluntarily or in response to prodding by the EPA's defunct Office of Noise Abatement Control—to produce safer, quieter gardening machines, it would still take decades for the older, noisier machines, some thirty-four million of them, to be retired, since the average life of a mower is seven years.

It is difficult for consumers to make informed decisions about gardening-machine purchases because manufacturers are not required to report their noise levels or emissions, even though leaf blowers, weed wackers, lawn mowers, and the like emit solid particulate matter and raise dust (entrain) from the ground. The entrain may contain lead, carbon, arsenic, cadmium, chromium, nickel, and mercury. The machines also stir up molds and pollens, known irritants to sufferers of asthma, as well as animal feces and the pesticides and herbicides we copiously apply to our gardens and lawns.[17] You have heard these machines, smelled them, and inhaled what they raise into the air. What health and safety concerns do scores of millions of such machines suggest?

While lawn and gardening machines have become quite popular for their ease of use and their power to transform unkempt spaces into kempt ones, many users of the machines have come to recognize the environmental and public health risks that come with their use. These concerns have prompted regulatory action at the local level, since officials at the state and federal level have been slow to act. One of the major actions has been to restrict the use of leaf blowers and weed wackers to certain hours of the day and to set decibel limits at a level lower than federal regulations permit. While widely accepted by weekend gardeners, the noisy machines have become less and less welcome as armies of gardening crews have begun to use them. The crews employ poorly trained workers, often Hispanic, many of them illegal aliens, to fan out through cities to attack leaves, dirt, dust, and other debris with their high-pitched machines.[18] The concentrations of noise and dust have upset many residents. Carmel-by-the-Sea, California, banned leaf blowers in 1975, and Beverly Hills followed suit in 1978. Twenty California cities have since banned leaf blowers, and at least eighty other cities have ordinances on the books restricting use, noise levels, or both. In July 2001, joining dozens of other municipalities in Canada and the United States, the city of Vancouver, British Columbia, adopted a law that lowered the acceptable decibel level of leaf blowers operated within city limits, further stipulating that they be run at the lowest effective throttle setting. Officials banned their use on Sundays and holidays and limited their hours of operation to between 8 A.M. and 6 P.M. on weekdays and to between 9 and 5 on Saturdays. Citizens' complaints had identified leaf blowers as "one of the most readily identifiable sources of annoyance," third after construction noise and loud music. Leaf blowers had become "acoustic public enemy number one."[19] Legislators in Arizona and New Jersey have considered laws regulating leaf blowers.

Opponents of regulation have raised the issue of racial discrimination. They point out that minority populations are often hired to perform domestic lawn work, and that limiting the use of leaf blowers and weed wackers will cut into jobs and have a disproportionate impact on groups who can least afford it. Landscaping companies in California, Arizona, and Texas, where many of the workers are Mexicans or Mexican-Americans, oppose regulations on the use of the machines, claiming that they eliminate jobs and so deny the children of migrant workers the finances to go to college. Yet the owners of these companies rarely offer their workers eye or ear protection, let alone health insurance or pension programs. If they were truly concerned about workers and their children, they would provide those fringe benefits and require safety equipment.

As is fairly standard among purveyors of small-bore engines, gardening firms, municipalities, manufacturers, and others cite economics as the major reason leaf blowers should be regulated for safety or emissions only cautiously, if at all, lest they become too expensive. Old-fashioned brooms and rakes may be environmentally friendly, they acknowledge, but are too slow and labor intensive. The business of lawn maintenance demands the fastest work possible in order to do as many lawns as possible. Yet the fact that low-tech leaf blowers could be cited "as essential to the livelihoods of thousands of people in Dallas, Chicago, and Los Angeles reveals a far greater problem than efficiency, aesthetics, and pollution. At a time when the skills and educational requirements for well-paying jobs are higher than ever before, the supply of functionally illiterate workers continues to grow."[20] Couldn't lawn companies employ more workers and create a few more jobs while cutting costs for oil, gas, and machines if they used rakes and brooms?

The California Landscape Contractors Association (CLCA), one of the more progressive user organizations in the battle over gasoline-powered gardening machines, has sought to portray motorized gardening equipment as extremely efficient and safe. The membership opposes across-the-board bans against machines as unnecessary, bad public policy, and harmful to the industry. They see the tools as essential for efficiently cleaning up small debris and successfully supplanting brooms, rakes, and hoses while working such surfaces as rock, gravel, and mulch that rakes cannot. The devices save time and labor, CLCA members say; similar work done without the machines would cost 20 percent more, a cost that would hurt the middle class, the poor, and the elderly, who would have to watch as their landscapes "deteriorated." They point out that power equipment saves water and meets federal standards. And if some machines are a nuisance, the culprits are old technology and improper use.[21] But haven't we heard all of these arguments elsewhere with respect to other machines? It's never the machine, it's the bad user. It's never the machine, it's the outdated design. And, the argument goes, only the poor, the weak, the meek, and the Hispanic will suffer when machines are banned.[22] Yet read any advertisement from a manufacturer or from the CLCA, and you will never see a mention of pollution or public health.

In the effort to curb small-bore engines as a significant source of pollution, in 1996 the EPA introduced a standard whereby manufacturers of engines of 25 hp or less would be required to cut emissions of carbon monoxide, hydrocarbons, nitrous oxides, and other pollutants. The cost of improvement might be $5 to $7 per engine, but efficiency and reliability would increase, resulting in lower operating costs. The EPA mounted a public relations campaign to involve equipment owners in the effort to lower emissions. It urged weekend gardeners to follow a series of easy steps to cut pollution: avoid spilling fuel, maintain equipment, consider such "cleaner" options as electric devices and manual tools, decrease the size of lawns, plant types of grass that require less mowing.[23]

Whirling Amputations

Regarding health and safety, anecdotal evidence and formal data indicate that Americans do not believe their gardening machines to be dangerous or that they overestimate their ability to use them properly. Each year an average of 132 American farm workers are crushed to death when tractors overturn during operation. Officials at the National Institute for Occupational Safety and Health (NIOSH) believe that virtually all of these deaths could have been prevented. In 1993, then–NIOSH director J. Donald Millar called tractor rollovers an "occupational obscenity." According to Millar, "There is no scientific excuse for the persistence of this problem. This is something we know how to prevent." The key to prevention is the presence of a rollover protective structure (ROPS) on every tractor in use. What Millar means is something like a roll bar, and the structure can be either enclosed or open. Operators would have to use such safety devices as seatbelts.[24]

What of death and injury around the home? Lawn mower injuries became a source of concern in the late 1970s and early 1980s. This led to a federal standard issued in 1982 that required a rotary blade to stop within three seconds after the user left the operator position at the rear of the mower.[25] The Consumer Product Safety Commission (CPSC) followed in 1987 with a mandatory safety standard to reduce the blade-contact hazard of walk-behind power lawn mowers. About 50 percent of all lawn mower injuries and 64 percent of walk-behind mower injuries occurred as a result of contact with the moving blade, amounting to seventy-seven thousand blade-contact injuries annually. The standard would eliminate the vast majority of these injuries, CPSC officials claimed, saving $211 million (in 1987 dollars) annually in treatment costs (not including the costs of pain and suffering). The standard would add $35 to the price of a typical lawn mower.[26] Such standards were opposed by some manufacturers as too costly, but the number of toes and fingers separated from their owners by spinning blades indicated that action was necessary.

In the same way that some farmers took for granted the awesome power of a tractor to inflict injury, some consumers failed to understand that life and limb are at risk when gardening around the home. Yet lawn mowers cause 83 percent of foot amputations to children, according to a November 1996 *Journal of Trauma* article. More than sixty thousand Americans bleeding from injuries related to lawn-mowing find their way into hospital emergency rooms every year. Lawn mowers cause nearly half (46%) of all injuries to children under the age of five and a third (34%) of all pediatric amputations. In a fit of understatement Jerry Jurkovich, one of the trauma specialists involved in the study, said, "Our research shows that lawn mowers can be extremely dangerous." Jurkovich noted that children under the age of fifteen have more lawn mower injuries than any other age group. He recommends that only properly trained children fifteen years old or older should be allowed to mow the lawn alone.[27]

The more machines, the more injuries. Chain saws inflict their own sorts of severe trauma owing to the phenomenon of "kickback," in which the saw, its chain blades moving at high speed, suddenly rears up and back toward the user. The effort to develop an industrywide safety standard picked up in the mid-1970s with an increase in accident and death rates. The standard, agreed to by the Chain Saw Manufacturers Association, was to be voluntary. Between 1973 and 1978 there were thirty deaths linked to chain saws, most of them occurring when the blades of the saw severed the blood vessels of the neck. In 1977 almost forty thousand people with chain saw injuries were treated in emergency rooms, a 47 percent increase over 1975.[28] The number of chain saw injuries serious enough to require emergency room care increased from twenty-nine thousand in 1976 to fifty-three thousand in 1979. Roughly three million chain saws were being sold annually in the late 1970s, raising the specter of even larger numbers of serious injuries, which prompted the CPSC to require the development of low-kickback chain saws equipped with nose tip guards, low-kick guide bars, chain brakes, hand guards, and so-called old-technology low-kick chains. Because the industry quickly adopted a voluntary standard to reduce kickback, the CPSC voted to terminate further work on a mandatory standard in 1985. The CPSC reported that by 1982 the impact of kickback had been reduced to twenty-two thousand medically attended chain saw injuries annually. The CPSC urged consumers to retrofit older saws with low-kickback replacement chains.[29]

This experience indicates that the pursuit of voluntary standards can achieve meaningful regulations, improvements in technology, and reductions in injuries. Industry, government, and citizens' groups essentially seek the same ends. But government pressure and guidance through setting standards and using the courts to enforce compliance seem to be crucial ingredients to ensure improvements in machines with small-bore engines, machines that otherwise will continue to have great social and environmental costs.

Gardening machines, like their recreational-machine cousins, operate at high speeds using powerful engines. At high power, they vibrate and hum. These properties mean that they must be used with great care, and that they require regular maintenance and repair. Manufacturers of gardening equipment have had to recall millions of the machines over the last twenty-five years alone, which indicates that even with proper maintenance, these machines are a potential source of danger to the public. For example, in April 1997 the CPSC and Husqvarna announced the recall of 277,000 chain saws because heat from the saw's muffler could melt the front hand guard, which was designed to prevent contact with the chain guard.[30]

Recreational and Gardening Machines at Home, Work, and Play

As Henry Ford was to the automobile, Briggs and Stratton, Tecumseh, and other small-engine manufacturers have been to leaf blowers, lawn mowers, and recreational machines. The internal combustion engine is the epitome of American technology. We mass-produce it. We've taught others to mass-produce it. It moves us, powers us, stocks our stores. Owning one is the dream of most teenagers. The internal combustion engine is not merely a device, it is an institution, a series of interconnected and mutually dependent systems. These systems include manufacturers and the trade associations that represent them; oil companies and road construction firms; the governmental departments and agencies that regulate the engines; the boating, ATV, and snowmobile clubs whose members band together to defend their right to use the mighty gasoline-powered engine, to secure open waters and well-maintained trails through state and federal lands; the after-market companies that sell engine, shock, and other machine modification equipment, insignia shirts, decals, books, gloves, and helmets; and the hospitals where too many users end up. Whether lawn mowers, weed wackers, snowmobiles, jet skis, or ATVs, the engines provide loads of power, usually more than most users require, and produce lots of noise, lots of pollution, and, often, lots of fun. Precisely because a recreational machine consists of a series of interconnected and interdependent systems, we have failed to integrate it into society properly. We see only the machine itself, not its attendant systems and therefore not its significant social and environmental costs.

I live in the Kennebec River Valley of central Maine. The Abnaki Indians lived in the region for thousands of years until settlers, French and English speaking, spread through the region. They found in central Maine rich hunting grounds of moose, deer, bear, and many smaller mammals; plentiful salmon, trout, and even sturgeon in the region's rivers and lakes; and dense, magnificent forests of pine, spruce, and fir, some of the trees measuring 6 feet wide at the base and stretching 150 feet straight up into the sky. The trees gave rise to various markets, including shipbuilding, construction materials such as planks and shingles, and, by the end of the nineteenth century, paper. Mill towns appeared up and down the rivers, including Waterville, where I live. Many of the mill towns are in crisis, and some are dying as industry moves away. And the forests that incredulous settlers first smelled 70 miles out into the Atlantic on their approach to North America are now in their third and fourth growths. None of the trees is as old or tall as those the lumbermen harvested just 150 years ago. Yet Maine remains 90 percent forest, a green if not an evergreen state. Now everywhere among those trees in those forests are tens of thousands of snowmobiles and ATVs.

So when I head out for my morning run, I smell the conifers too, and I can be among them within forty or fifty minutes. The vistas from the top of the valley are calming, whether in winter or summer, but I remain on edge, for I never know when I will encounter the next snowmobile or ATV. I can smell and hear them all too frequently. The views have been disfigured by muddy, rutted trails that grow wider and deeper each season. I am confused by the fact that a town undergoing economic hardship in a state that is watching jobs go south and abroad can support the hundreds of ATVs, snowmobiles, other machines, with their trailers and clothing and boots and helmets, at thousands of dollars for each complete machine package.

I am continually amazed that so many operators seem oblivious to the inherent dangers of these machines or overestimate their ability to handle them at high speeds and on uncertain terrain. Perhaps we should allow Darwinian laws of natural selection to play out as one foolish operator after another kills himself. But should we not at least protect their children? One early fall day as I chugged up the hill on Gagnon Road in Oakland, I stopped dead in my tracks at the sight of a man operating a ride-on mower with a three-year-old child sitting on his lap. Both were barefoot and in shorts; neither wore a helmet or eye or ear protection. The man was steering the 300-pound machine with one hand while holding the boy with the other. If he knew that over eighty thousand people are hospitalized annually with lawn mower injuries, including amputations, he ignored that fact. I often see children under the age of sixteen driving ATVs of various sorts, against the law, against common sense, against traffic, in places where they are banned, on machines that were outlawed in 1988 but continue to be sold secondhand legally. Those children never wear helmets.

Recreation, hunting, and fishing should be a part of the activities that occur in the forest, in the park, and elsewhere. But the ubiquity of recreational and gardening machines, and their potential risks and costs, should make us ask, To what extent can recreation, hunting, and fishing be compatible with the millions of recreational machines already in the forest, on the dunes, in the water, on the ice and snow, with millions more certain to be bought? Their noise, their smell, their pollution, and the way they injure, maim, and kill people, even when their operators follow the manuals closely, have convinced me that we must reconsider their place. I have talked with owners of the stores that sell the machines; with manufacturers and their trade association representatives; and with people at the factories that build the machines. They are all good people. But this does not mean that we should settle for regulation by the regulated, or allow children to use these machines, or to permit adults to use them without helmets and other safety equipment, or to cease pushing for improvements in safety and emissions controls. On the contrary. I do not believe that the U.S. Constitution—which, by the way, never mentions the small-bore engine—guarantees the right to unregulated Fordist recreation, for engine-powered recreation is a public health and environmental danger to us all.

There are many risks to living in modern industrial society, and most Americans accept them. Why not? After all, they live well, their lives are long, their food is good, their medical technology is the world's best, and they have established laws to clean up hazardous waste, limit air and water pollution, and protect wilderness areas. If they knew the great costs of internal combustion engines, perhaps they would call for greater regulation of them, too. But because of their convenience, their time- and labor-saving qualities, and the access they give us to faraway places, we have embraced internal combustion engines in a variety of ecosystems, often without thinking through the costs: thousands of deaths; hundreds of thousands of injuries to operators and bystanders alike, ranging from paralysis to loss of hearing, sight, limbs, and fingers; the destruction of flora, fauna, and the ecosystems of which they are a part; air and water pollution.

Some of the recreational machines that use internal combustion engines remain inherently unsafe to operate even after twenty years of industry-promoted, federally mandated, or commonsense improvements. The two machines that virtually define "inherently unsafe" are the ATV and the jet ski. The ATV was originally designed with three wheels, most likely because three-wheelers were cheaper to produce and seemed stable enough. Unfortunately, three-wheeled ATVs tended to flip over even at low speeds. It is now illegal to manufacture three-wheelers, but many remain on the resale market. Adding another tire and widening the distance between them on each axle helped reduce the tip-over problem somewhat. But given the speeds they can reach, the terrain their operators tend to explore, and their high center of gravity, even four-wheeled ATVs can flip or roll over, if less frequently than earlier models. While accident rates have declined, the number of ATVs has increased rapidly, resulting in a fairly constant number of deaths, probably five hundred annually nationwide. Yet any mention of regulation to improve the safety of such vehicles raises concern that government officials are trampling individual rights and denying life, liberty, and the pursuit of happiness. These arguments are familiar to those who have studied the difficulty of passing seatbelt and motorcycle helmet laws, and they remain a central feature of the American mind-set with respect to machines.

Similarly, the use of jet skis has led to hundreds of deaths (from blunt trauma injuries, not drowning) and thousands of hospitalizations annually. These deaths and injuries occur in part because operators fail to take safety courses (recommended but not required in most states), in part because they do not read the owners' manuals, in part because alcohol often figures in accidents (as it does with accidents involving heavy machinery of any sort), and in part because the technology is inherently unsafe. In this case the problem arises because jet skis maneuver poorly when the rider naturally lets go of the throttle mechanism during an emergency maneuver. More precisely, it is counterintuitive to maintain speed during emergency maneuvers. But when you slow a jet ski by releasing the throttle handle, you lose the ability to steer it, and it continues going in the direction you are trying to avoid. So-called off-throttle steering mechanisms may reduce accidents.

What about you? Try swimming in your favorite lakeside park. The odor of two-stroke exhaust fills the air, a film of gas and oil coats the surface, and the choppy waves produced by powerboats and jet skis make swimming unpleasant. Swimmers pick up that film on their skin and sometimes unintentionally swallow a gulp. Things are no better in the winter. Snowmobiles have tens of thousands of miles of trails through the woods, many of them state supported, many of them club maintained, some of them illegal. Of course, fast is fast, and too many operators drive with excessive speed. This means that they do not share the trails with cross-country skiers, they endanger them. And the snow vehicles themselves give off noxious smoke, their noise frightens animals, and they kill their operators. In Michigan they kill an average of thirty operators annually. In Maine, New Hampshire, and Vermont, the winter of 2002–3 took twenty-eight snowmobilers to their death. If the rights of snowmobilers to use snowmobiles are an important consideration, so are the rights of their families to see them come home in one piece at the end of the day, and so are the rights of other recreationists who wish for a more peaceful recreational experience.

Manifest Destiny: Combustophilia

I have a friend who owns seventeen internal combustion engines: three automobiles, one Jeep, two weed wackers, two lawn mowers, two boat motors, two chain saws, two generators, one water pump, and one "personal" hovercraft, itself having two engines. He intends no injustice to the environment. He is but an American homeowner who considers these engines the very thing to help him rightfully enjoy his yard, his garden, and his trips to the countryside. Yet they seem out of place next to his solar-powered laptop computer. How did America become the land of the internal combustion engine? Why do Americans have such fierce pride in their recreational and gardening machines? Where did feelings of the manifest right to conquer nature with engines originate?

The phenomenon of "combustophilia" has evolved largely in parallel with that of "automobility." A central aspect of combustophilia is Americans' belief that it is their right to conquer nature, a notion many historians and philosophers attribute to ideas current even in the early days of the republic. Yet few people have considered the persistence of this idea into the twenty-first century or its evidence throughout popular culture (in club memberships, T-shirts, advertising, and so on) and in material culture (in this case, recreational and gardening machines). The Fordist revolution in the mass production of automobiles, the growth of disposable income, and the increase in leisure time all made machines of all sorts accessible to postwar middle-income Americans—and to workers, too. The average American homeowner sports several internal combustion engines, a garage to house them, and a trailer to haul them. In the same way that automobiles contributed to changes in American lifestyle in courting practices, purchasing behaviors, and housing patterns, so combustophilia represents changed behaviors for starting and finishing each day, spending weekends in the yard, and keeping machines in good operating condition. Americans have come to feel comfortable lopping the head off a weed so inefficiently with a 3-hp motor, so incongruously experiencing what they call the delights of the garden, the woods, the parklands, the lakes and rivers, while wielding a powerful cutter, shredder, or blower or mounted atop a hot, smoky machine.

Rachel Carson warned us forty years ago:

> Only within the moment of time represented by the present century has one species—man—acquired significant power to alter the nature of his world. During the past quarter century this power has not only increased to one of disturbing magnitude but it has changed in character. The most alarming of all man's assaults upon the environment is the contamination of air, earth, rivers, and sea with dangerous and even lethal materials. This pollution is for the most part irrecoverable; the chain of evil it initiates not only in the world that must support life but in living tissues is for the most part irreversible.[31]

No less than the dangers of chemicals, we must actively consider the dangers of recreational machines, understand how their use leads to disturbing changes in ecosystems, and recognize our duty and ability to do something about it.

In the early 1990s I bought a house in New Hampshire with a large yard. Like all good American homeowners, I then purchased a lawn mower and cut the lawn. Ten days later I cut it again. I had already learned to hate the dust, dirt, and noise, even though I wore leather gloves and protective eyewear and earphones. After the third cutting I began to question my sanity. I had spent an hour and a half pushing a 21-inch mower powered by a 4-hp Briggs and Stratton engine over scraggly grass and patches of weeds, breathing noxious fumes and unavoidably sucking toads into the mulcher. My investigations revealed that the previous owner had spent many of his waking hours pondering how to produce a green carpet of monocultural grass. He had spent hundreds of dollars on the failed attempt. I vowed I would not make the same mistake. In the name of abating air and noise pollution, ending the use of chemical pesticides, herbicides, and fertilizers, and protecting my own time and lungs, I quit cutting the front lawn, cold turkey. It was, to be sure, the northern side of the house, where a lawn in this hemisphere might never grow well in any event. The backyard, where children played soccer, football, and baseball, met the lawn mower from time to time. But I stopped watering it or applying chemicals to it, and I cut it no more than a half-dozen times a year, mulching the clippings back into the soil.

The front yard required more thought. Then I came up with the solution: I abandoned the internal combustion engine. I went to my neighbors, many of whom had forested areas abutting their properties, and asked permission to dig up birch, maple, oak, spruce, fir, and pine saplings from these areas. Some were a foot high, some 5 feet high. I planted year round, but especially in late winter and late fall. I bought trees through catalogues, a dozen pine for $8 here, ten poplar for $10 there. And I planted. The front yard—165 feet wide and 60 feet deep from the road to the house, and bounded by bushes and trees from the neighbors' houses—had been filled with the scraggliest of grass. Within a year, wild strawberries and black-eyed susans had appeared from nowhere. Garter snakes, toads, and other signs of ecosystem vitality joined the front yard community. To the consternation of neighbors and the town's snowplow operators, I brought in 17 tons of stone and over two weeks built a wall to separate the yard from the street. The trees grew and grew and grew, and forest primeval now obscures the house. I had liberated myself from the lawn mower. When I moved to central Maine and into another house, I commenced the same project. My front yard has nearly been liberated from the lawn mower; birch, spruce, and maples have begun to shoot up; toads and snakes have reappeared.

We see roadkill—raccoons, possums, porcupines, squirrels—from our automobiles as we blaze down the highway, we are admonished about the dangers of moose and deer collisions, yet we seem oblivious to the carnage we inflict on animals small and large when we use recreational vehicles. For example, many pond-breeding amphibians must cross roads to reach breeding, summering, or hibernation sites. A Canadian herpetologist investigated a 20-km stretch of secondary road within a national park of eastern Canada over a seven-year period. He observed—a bit tautologically—an increasing number of dead toads with increasing traffic intensity. The scientist suggested careful study of wetlands and other habitat in all circumstances before any road building commenced.[32] All studies of the impact on animals of snowmobiles, ATVs, and other off-road vehicles indicate that the impact is great, and that it is growing with the numbers of vehicles.

Today's recreational vehicles do not resemble their forebears in terms of noise and air pollution. Both quieter and more efficient, in numbers of one they intrude less on solitude and ecosystems. But there are millions of them, and the numbers are growing, and millions of internal combustion engines produce noise and emissions. They have found their way into parks before adequate attention has been paid to their effects on public health and the environment. This is insufficient reason to leave them there or permit more of them. They must be restricted, or our children will have little resembling what their parents had in which to recreate.

Local people with local issues are the backbone of the recreational- and gardening-machine industries. Their feelings and beliefs about ATVs, jet skis, and snowmobiles are crucial to understanding how these machines have taken root in America. This has been the story of their machines. Let's help these recreationists continue to use them while allowing other people to enjoy recreation without the use of machines, if they so choose. Let's urge Americans—in the name of life, liberty, and a clean environment—to abandon their devotion to lawns and lawn mowers.

And let's get rid of unsafe recreational machines. Lower the center of gravity, build roll bars, add seatbelts, power down those engines. Treat them like automobiles, and prohibit children under sixteen from using them.

Remember that it does no one any good for the U.S. Department of the Interior, the Bureau of Land Management, and the National Park Service to abdicate their responsibility to protect federal lands from the extensive damage caused by recreational machines because of past practices that admitted motorized vehicles before their impact became clear and before there were so many of them. Congress and state governments must provide land managers with the resources to study the situation and to enforce the laws.

And finally, if you insist upon using recreational and gardening machines, wear safety equipment, be sensitive to the needs of others, worry about wildlife and ecosystems, don't make any new trails, and pray you don't hit something or fall off.

APPENDIX

Table A.1. Registered Snowmobiles by State, 2005–2006

Alaska	53,593	Nebraska	2,450
Arizona	n.a.	New Hampshire	44,000
California	22,330	New York	149,610
Colorado	36,500	North Dakota	18,185
Idaho	48,900	Ohio	17,300
Illinois	41,897	Oregon	17,092
Indiana	13,499	Pennsylvania	46,564
Iowa	40,650	South Dakota	11,898
Maine	73,275	Utah	28,221
Massachusetts	19,000	Vermont	34,743
Michigan	374,522	Washington	35,500
Minnesota	278,886	Wisconsin	215,758
Montana	31,259	Wyoming	34,852
		Total	1,690,484

Source: www.snowmobile.org/stats_registrations_us.asp.

Table A.2. Worldwide Snowmobile Sales, 1992–2006

1992	150,000	2000	208,297
1993	158,000	2001	208,592
1994	181,000	2002	203,153
1995	227,443	2003	186,627
1996	252,324	2004	181,336
1997	260,735	2005	173,733
1998	257,936	2006	164,860
1999	230,887		

Source: http://72.14.203.104/search?q=cache:2TVVIYLMXCcJ:www.snowmobile.org/stats_sales_worldwide.asp+%22snowmobile+sales%22%2B1992&hl=en&gl=us&ct=clnk&cd=5.

Table A.3. Miles of Groomed Snowmobile Trails by State, 2005–2006

Alaska	350	Nebraska	404
Arizona	500	New Hampshire	7,000
California/Nevada	2,500	New York	10,674
Colorado	2,800	North Dakota	3,650
Idaho	7,200	Ohio	150
Illinois	2,500	Oregon	6,410
Indiana	210	Pennsylvania	3,363
Iowa	5,000	South Dakota	1,572
Maine	13,500	Utah	1,190
Massachusetts	1,100	Vermont	4,670
Michigan	6,260	Washington	3,000
Minnesota	20,385	Wisconsin	19,099
Montana	4,071	Wyoming	2,400
		Total	129,958

Source: www.snowmobile.org/facts_snowtrails.asp.

Table A.4. ATV-Related Deaths and Injuries, 1982–2001

	Reported Deaths	Emergency Room–Treated Injuries
1982	29	10,100
1983	85	32,100
1984	156	77,900
1985	251	105,700
1986	299	106,000
1987	264	93,600
1988	250	74,600
1989	230	70,300
1990	234	59,500
1991	230	58,100
1992	221	58,200
1993	183	49,800
1994	198	50,800
1995	200	52,200
1996	248	53,600
1997	241	54,700
1998	251	70,200

continued

Table A.4. *continued*

	Reported Deaths	Emergency Room–Treated Injuries
1999	357	85,100
2000	344	95,500
2001	270	111,700

Source: Robin Ingle, *Annual Report of ATV Deaths and Injuries* (Washington, DC: Consumer Product Safety Commission, 2002).

Note: Based on data generated from CPSC's National Electronic Injury Surveillance System, estimated deaths are 20–40% higher than reported deaths, depending on the year.

Table A.5. Impacts on Wildlife Linked to Recreational Boating

Impact	Example
Alarm or flight	Nest flushing, rookery evacuation
Avoidance or displacement	Nest abandonment, migration disruption
Behavioral alteration	Decreased foraging or feeding
Community alteration	Increased predation
Habitat loss	Seagrass destruction, shoreline erosion
Injury or death	Vessel collisions, sediment-related gill damage
Reproductive failure	Decreased mating, increased egg mortality

Source: Impacts of Recreational Boating and PWC Use, chap. 2, Commonwealth of Massachusetts PWC Management Guide, at http://64.233.187.104/search?q=cache:NrUhtzTAjgIJ:www.mass.gov/czm/pwcmgntguide2.pdf+impacts+of+recreational+boating+and+pwc+use&hl=en, 31.

NOTES

Chapter 1. Fordism in Outdoor Recreation

1. "2006 Dinli Cobia 50," *ATV 4-Wheel Action,* September 2005, 93–95.

2. For a detailed historical sketch of the early developments, see Dugald Clerk, *The Gas, Petrol, and Oil Engine,* vol. 1 (New York: John Wiley and Sons, 1909), 1–50. See also Rolla Carpenter and H. Diederichs, *Internal Combustion Engines* (New York: D. Van Nostrand Company, 1908), 232–62.

3. On the history of the earthmoving equipment industry, see William Haycraft, *Yellow Steel* (Urbana: University of Illinois Press, 2000).

4. On Fordson tractors, see Robert Williams, *Fordson, Farmall, and Poppin' Johnny: A History of the Farm Tractor and Its Impact on America* (Urbana: University of Illinois Press, 1987).

5. Fred Jones, *Farm Gas Engines and Tractors* (New York: McGraw-Hill Book Company, 1932), 253–60.

6. www.acbs-bslol.com/Porthole/OleEvinrudePT2.htm. Another entrant to the small engine business was T&M. In 1892 John Termatt and Louis Monahan formed a partnership to build gas engines. They sold the business in 1902, in 1903 organizing another engine company called T&M, focusing on small marine engines. These proved very successful, and they built larger and larger ones using multiple cylinders, up to a 190-hp four-cylinder model. By 1906 the T&M Company was building small engine–driven generators for farm and home use, and in 1912 the company produced its first four-stroke engine, with a 2½-inch bore and 3½-inch stroke. Their engines found applications for the U.S. Army in World War I in generators to serve as pumps, some of which were employed in the Panama Canal and several of which Admiral Byrd used on an expedition to the South Pole. In 1914 T&M organized the marine engine business as the Universal Motor Company, while T&M continued to build farm-type engines in a variety of sizes from 1 to 12 hp. While the T&M farm engine business seems to have ended in 1925, Universal became the number one builder of auxiliary power plants for sailboats in the world and has survived in various incarnations to the present. See www.oldmarineengine.com/history/Termatt%20Monahan/TermattCoHistory.html.

7. www.mercurymarine.com/company_history.

8. www.yamaha-motor.co.jp/global/product-history/pp/engine/p7a-pc3/.

9. Clerk, *Gas, Petrol, and Oil Engine,* 1:321–27. Clerk put two cylinders side by side for higher efficiency and power than early one-cylinder units. By 1898 the American Society of Mechanical Engineers had already established a committee to test gas engines, their specifications, efficiencies, speed, fuel consumption, and so on. See Carpenter and Diederichs, *Internal Combustion Engines,* 486.

10. A. R. Rogowski, *Elements of Internal Combustion Engines* (New York: McGraw-Hill, 1953), 200–205; Carpenter and Diederichs, *Internal Combustion Engines,* 100–110, 364–66.

11. C. Fayette Taylor and Edward S. Taylor, *The Internal-Combustion Engine,* 2nd ed. (Scranton, PA: International Textbook Company, 1961), 272–75. One author noted that two-stroke engines might be difficult to start and operate "due largely to the fact that complete exhaust of the burned fuel residue is extremely difficult. Likewise, the problem of producing the correct fuel mixture and placing it in the cylinder is a difficult one. Unless such an engine is in perfect condition and correctly adjusted at all times, trouble will be apt to develop." See Jones, *Farm Gas Engines and Tractors,* 33–34.

12. According to engine specialists writing in 1913, "In small gasoline engines operating on the two-stroke principle with crank-case compression the lubricating oil is sometimes mixed with the gasoline and fed into the engine through the carburetor. Part of it is left suspended in the air as a fine oil fog after the gasoline vaporizes and this fog is carried by the mixture through the crank case and into the cylinder." C. F. Hirshfeld and T. C. Ulbricht, *Gas Engines for the Farm* (New York: John Wiley and Sons, Inc., 1913), 121–22.

13. A. H. Goldingham, *The Design and Construction of Oil Engines,* 5th ed. (New York: Spon and Chamberlain, 1922), 13, 66, 72–74, 85–86.

14. *Impacts of Recreational Boating and PWC Use,* chap. 2, Commonwealth of Massachusetts PWC Management Guide, at http://64.233.187.104/search?q=cache:NrUhtzTAjgIJ:www.mass.gov/czm/pwcmgntguide2.pdf+impacts+of+recreational+boating+and+pwc+use&hl=en, 22–27. MBTE is not biodegradable, and it tends to build up in aquatic areas. In spite of growing awareness of its environmental and other costs, MBTE had such strong protectors in the U.S. Congress as former majority leader Tom Delay, who sought legislation requiring taxpayers to pay the cost of clean-up and protecting MBTE manufacturers, including subsidiaries of Saudi producers, from liability.

15. Paul Burman and Frank DeLuca, *Fuel Injection and Controls for Internal Combustion Engines* (New York: Simmons-Boardman, 1962), 1–8.

16. "Program Update for Off-Road Motorcycles and ATVs," October 22, 1998, at www.arb.ca.gov/msprog/offroad/mcfactst.htm.

17. Ibid.

18. Juliet Eilperin, "Lawnmower Smog Rule Delayed," *Washington Post,* June 10, 2005, A21, at www.washingtonpost.com/wp-dyn/content/article/2005/06/09/AR2005060901774.html.

19. Ibid.

20. www.nonoise.org/aboutno.htm. See also George Maling Jr., "An Editor's View of the EPA Noise Program," *Noise Control Engineering Journal* 51, no. 3 (2003): 143–49. The Institute of Noise Control Engineering serves as a major clearinghouse for noise studies. G. Janssen and P. de Vos, "On Influencing Noise Legislation and Noise-Abatement Policy," *Journal of Sound and Vibration* 231, no. 3 (2000): 951–59.

21. One solution to road noise is a barrier of vegetation, wood, or concrete. In industrial settings there are a variety of ways to quiet machines. See, for example, R. E. Fairfax, "Noise Abatement Techniques in Southern Pine Sawmills and Planer Mills," *American Industrial Hygiene Association Journal* 50, no. 12 (1989): 634–38.

22. Sidney A. Shapiro, "The Dormant Noise Control Act and Options to Abate Noise Pollution," at www.nonoise.org/library/shapiro/shapiro.htm. See also Shapiro, "Lessons from a Public Policy Failure: EPA and Noise Abatement," *Ecology Law Quarterly* 19, no. 1 (1992): 1–61.

23. Shapiro, "Dormant Noise Control Act"; Shapiro, "Lessons from a Public Policy Failure."

24. Shapiro, "Dormant Noise Control Act."

25. Environmental Protection Agency, Office of Noise Abatement Control, *Noise: A Health Problem* (Washington, DC: EPA, August 1978). See also EPA, *Public Health and Welfare Criteria for Noise* (Washington, DC: EPA, July 27, 1973), and W. Vermeer-Passchier and W. F. Passchier, "Noise Exposure and Public Health," *Environmental Health Perspectives* 108, suppl. 1 (March 2000): 123–31. In all spheres of life Americans are exposed to excessive noise. A study of midwestern farmers focused on all of the ways in which they were exposed: farming equipment operation, chain saw use, welding and metal work, handling of large animals in and out of confinement facilities, feed handling, manure storage facilities, and mixing and applying chemicals. With the exception of welding masks, the use of personal protective equipment was low, although public health officials called for such equipment to be used to reduce the risk of personal injury or exposure. W. S. Carpenter et al., "Assessment of Personal Protective Equipment Use among Midwestern Farmers," *American Journal of Industrial Medicine* 42, no. 3 (2002): 236–47.

26. www.atvaonline.com/news/03/califnoise.asp.

27. Wisconsin accepted the California standard to prohibit a person or manufacturer from selling, renting, or operating an ATV if the noise level exceeded 96 dB. In the same act, Wisconsin legislators required nonresident ATV users to display a nonresident trail place (unless exempted for municipalities, federal agencies, or American Indian tribes or bands). This was required because Wisconsin had become a weekend recreation spot, and drivers from states without a registration requirement (specifically Illinois and Michigan) arrived with vehicles whose sound characteristics exceeded 96 dB or rode in violation of the law. See http://66.102.7.104/search?q=cache:babkkHROOcMJ:www.dnr.state.wi.us/org/nrboard/minutes/M05/0505%2520minutes.pdf+decibels+atvs&hl=en.

28. www.nonoise.org/aboutno.htm.

29. "Snowmobiles and the Environment," *Yale Law Journal* 82, no. 4 (1973): 774, 779.

30. James M. Glover, *A Wilderness Original: The Life of Bob Marshall* (Seattle: Mountaineers, 1986), 168. The evidence supported Marshall's call for greater control of forest resources, although his reference to state control raised the specter of socialism in the view of his critics. They had reason to fear socialism, for they had seen the results of state exploitation of resources in Stalin's Russia. Yet Marshall also had deep reasons for concern about the health of the forests. In his *Alaska Wilderness* and *A People's Forests,* among other writings, he suggested that state control of forest resources was the only path, for forestry had become a profession that stood "brazenly for forest depletion." The Wilderness Society included among its original members Aldo Leopold, a pioneer of wildlife ecology known for his *Sand County Almanac,* and it continues its activities seventy years later to defend the last few wilderness areas of the United States from human encroachment.

31. Glover, *Wilderness Original,* 3, 168–69, 172, 175, 194–95.

32. Robert Marshall, "The Problem of Wilderness," *Scientific Monthly* 30, no. 2 (1930): 141–42.

33. Ibid., 143–45.

34. Ibid., 147.

35. Izaak Walton League, *Caught in the Treads: Unethical Advertising in the AV Industry* (Gaithersburg, MD: IWL, 2002).

36. Ibid.

37. Ibid.

38. *Tread Lightly Guide to Responsible Four Wheeling with Minimum Impact Camping Tips,* 3–6. The concept of "minimum impact camping" with an SUV is totally incongruous.

39. *Tread Lightly Guide to Responsible ATV Riding with Minimum Impact Camping Tips,* 1–18.

40. *Tread Lightly Guide to Responsible Snowmobiling,* 5–9. *Tread Lightly Guide to Responsible Personal Watercraft Use,* 2–7, 11.

41. *Tread Lightly Guide to Responsible Snowmobiling,* 12.

42. U.S. Forest Service, North American Association of Hunter Safety Coordination, *Tread Lightly* (Washington, DC: U.S. GPO, 1990).

43. Richard M. Nixon, Executive Order 11644, "On Use of Off-Road Vehicles on Public Lands," *Federal Register* 37, no. 27 (1972): 2877–78.

Chapter 2. Elk-Snowmobile Syndrome

1. Warren Upham, *The Glacial Lake Agassiz* (Washington, DC: U.S. GPO, 1895), 582.

2. Moose Utility, *Quality ATV Products* (2006).

3. www.cpsc.gov/CPSCPUB/PUBS/541.html.

4. Norma Jean Lodico, *Environmental Effects of Off-Road Vehicles: A Review of the Literature,* bibliography series no. 29 (U.S. Department of the Interior, Office of Library Services, September 1973), 1–10.

5. Ibid., 10–16.

6. Scott Creel et al., "Snowmobile Activity and Glucocorticoid Stress Responses in Wolves and Elk," *Conservation Biology* 16, no. 3 (2002): 809–14.

7. See Rudy Boonstra, "Coping with Changing Northern Environments: The Role of the Stress Axis in Birds and Mammals," *Integrative and Comparative Biology* 44, no. 2 (2004): 95–108.

8. www.mainerec.com/logger4.shtml; Richard Judd, *Aroostook: A Century of Logging in Northern Maine* (Orono: University of Maine Press, 1989), 186–90.

9. Steve Pierce, "First Tracks," *Iron Dog Tracks* 20, no. 4 (2005): 1; Gene Schnaser, "Pioneers of Snow Travel," *Snow Goer Magazine,* 1981.

10. Schnaser, "Pioneers of Snow Travel."

11. Pierce, "First Tracks," 1.

12. Ibid.

13. Schnaser, "Pioneers of Snow Travel."

14. www.eliason-snowmobile.com/; www.eliason-snowmobile.com/phase/phase2.htm; www.eliason-snowmobile.com/phase/phase3.htm. A group of Soviet officials also visited the FWD plant to test-drive the vehicles. They borrowed a machine gun, mounted it up front, and, while running the river, sprayed blanks at the river banks.

15. Gene Schnaser, "Pioneers of Snow Travel—Part II," *Snow Goer Magazine,* 1981, 34–38. In the 1930s, 1940s, and 1950s several tinkerers

also produced snow planes. Ray Vigue of Waterville, Maine, built about fifteen snow planes, some with a top speed of 80 mph, and later versions employing aircraft engines with up to 140 hp.

16. www.trailasled.com/newblood.php.

17. Ibid.

18. Ibid.

19. Ibid.

20. Ibid.

21. www.bombardiermuseum.com/en/content/jab/biographie1926_1938.htm.

22. Ibid.

23. Leonard S. Reich, "Ski-Dogs, Pol-Cats, and the Mechanization of Winter: The Development of Recreational Snowmobiling in North America," *Technology and Culture* 40, no. 3 (1999): 484–516, at 493–98.

24. www.bombardiermuseum.com/en/content/jab/biographie1939_1945.htm.

25. Ibid.

26. www.bombardiermuseum.com/en/content/jab/biographie1946_1948.htm.

27. www.bombardiermuseum.com/en/content/jab/biographie1949_1958.htm.

28. Ibid. Experiencing problems with bolts, rubber, and other components (like Henry Ford, who had sought to build his own Fordist rubber plantation in the Amazon rainforest, called "Fordlandia"), Bombardier created new vulcanization processes that enabled production of the continuous track his vehicles required and fabricated sprockets, bolts, and other components, small and large, on site.

29. Reich, "Ski-Dogs, Pol-Cats, and the Mechanization of Winter," 493–98.

30. www.bombardiermuseum.com/en/content/jab/biographie1959_1964.htm.

31. *1961 Ski-Doo Brochure,* from Bombardier Museum, Valcourt, Quebec, Canada.

32. *1963 Ski-Doo Brochure,* from Bombardier Museum.

33. *1968 Ski-Doo Brochure* and *1969 Ski-Doo Brochure,* from Bombardier Museum.

34. Ernol "Bud" Knapp with Gale Urbainzick, "Hus-Ski and Bolens Articulated Snowmobiles," typescript, Antique Snowmobile Club of America, n.d.

35. C. J. Ramstad, *Legend: Arctic Cat's First Quarter Century* (Deephaven, MN: PPM Books, 1987), 15–18.

36. Interview with Mitchell Johnson, Roseau, MN, November 2, 2005.

37. From Edgar Hetteen's biography, *Breaking Trail,* and information found in Bill Vint's book, *Warriors of Winter,* as posted at www.snowridermag.com/youngriders.html.

38. Ramstad, *Legend,* 19–29.

39. Johnson interview, November 2, 2005.

40. Ramstad, *Legend,* 30–51.

41. Ibid., 53–73. One of the Arctic Cats made in 1971 used a Wankel engine. The Wankel offered a quiet, vibration-free ride, creating less disturbance to livestock and neighbors. It had good mileage and met a 77-dB target, but it lacked acceleration and had problems with engine seals and so was abandoned. See George Wormley, "The Wankel Engine: It's Here in Snowmobiles," *Farm Journal* 96 (December 1972): 24.

42. Ramstad, *Legend,* 75–108.

43. "A Red-Hot Winter for Snowmobiles," *Business Week,* January 10, 1970, 34.

44. *Polaris Post,* various, 1960s.

45. Johnson interview, November 2, 2005.

46. Joe Rychetnik, "The Winter Games," *Time,* January 24, 1969, 50–51.

47. *Polaris Post,* various, 1960s.

48. Johnson interview, November 2, 2005.

49. P. K. Snoof, "1969 Snowmobile Buyer's Guide," *Field and Stream* 74 (October 1969): 69–73, 108. Other manufacturers in the postwar snowmobile business included Howard Schrader with his rear-engined Snowbug in Sudbury, Ontario; Bouchard with Moto Ski, which in 1962 produced ten machines in La Pocatière, Quebec; and Rejean Houle of Wickham, Quebec, who built twenty Skiroules in 1963. By the mid-1970s, the "sport-minded era of prosperity in the industry" had created nearly two hundred manufacturers. Pierce, "First Tracks," 1.

50. Ernol "Bud" Knapp, 55370 Russell Street, Cedar Springs, MI, "Snowmobile Manufacturers," typescript.

51. *Gates Snow Vehicle and Recreation Equipment Belt Guide,* 1970.

52. "Struggling to Cope without Snow," *Business Week,* February 19, 1980, 66. See also www.bombardiermuseum.com/en/content/motoneige/hautetbasdelindustrie.htm.

53. www.snowmobile.org/pr_snowfacts.asp. For registrations by state, see www.snowmobile.org/stats_registrations_us.asp. For U.S. and Canadian sales, see www.snowmobile.org/stats_units_us.asp and www.snowmobile.org/stats_units_canada.asp.

54. www.snowmobile.org/pr_snowfacts.asp.

55. President Dr. James D. Perry (1968), at www.ascoa.org/gogebic/_cover.htm.

56. www.ascoa.org/gogebic/_page4.htm; www.ascoa.org/gogebic/_phamplet.htm.

57. http://slc.engr.wisc.edu/organizations.html.

58. www.snowmobile.org/pr_applaud.asp.

59. M. Lund, "Snowmobiles: The Hot Rod of Winter Sports," *Holiday* 46 (December 1969): 43, 96.

60. Reich, "Ski-Dogs, Pol-Cats, and the Mechanization of Winter," 493–98.

61. "What Snowmobiles Do Is Open up the Winter World," *Sunset* 142 (February 1969): 46, 49, 51.

62. Lynn White Jr., "Technology Assessment from the Stance of a Medieval Historian," *American Historical Review* 79, no. 1 (1974): 1–13, and *Medieval Technology and Social Change* (London: Oxford University Press, 1974).

63. H. Nilsen, E. Utsi, and K. Bonaa, "Dietary and Nutrient Intake of a Sami Population Living in Traditional Reindeer Herding Areas in North Norway," *International Journal of Circumpolar Health* 58, no. 2 (1999): 120–33. See also P. J. Blind, "Lapp Society Yesterday and Today: A Comparison," *Arctic Medical Research* 45 (1987): 68.

64. Dave McCabe, "If You Give an Inuit a Snowmobile," at www.mcgill.ca/reporter/33/04/wenzel/. The collapse of the world market for sealskins also made the Inuit less reliant on hunting and more on southern contacts and goods made accessible through snowmobiles. For a history of the impact of the snowmobile on the Lapplanders, see Pelto Pertti, *The Snowmobile Revolution: Technology and Social Change in the Arctic* (Menlo Park, CA: Cummings, 1973).

65. McCabe, "If You Give an Inuit a Snowmobile."

66. www.gov.nu.ca/Nunavut/English/departments/commissioner/iceland.shtml.

67. Ibid.

68. L. Peplinski, "The Dogs of the Inuit: Companions in Survival," at www.fao.org/ag/againfo/resources/documents/WAR/war/W0613B/w0613b0m.htm.

69. Ibid. Granted, dog teams "provide far better insurance when darkness or weather conditions reduce visibility to nil. An experienced team will be able to find or stay on a trail in the worst weather conditions; the same is not true of the snowmobile. 'Quick trip out, long walk home' is often said of the snowmobile, which has the potential to break down or run out of fuel." Ultimately the advantage of snowmobiles was clear: a team of fourteen dogs may travel 130 kilometers in seventeen hours, with each dog requiring between 3,000 and 6,000 calories of

food, and with distemper and other diseases contributing to high mortality in some seasons.

70. St. Onge, Paul. (1996) "Transport et mobilité des résidants du village de Kangiqsualujjuaq: Le cas de la motoneige." M.A. thesis in geography, McGill University.

71. K. L. Capozza, "Arctic Natives Contemplate Technology's Impact on Culture, Environment," *Environmental News Network,* October 5, 2001. See also A. Pekkarinen, H. Anttonen, and J. Hassi, "Prevention of Accidents in Reindeer Herding Work," *Arctic Medical Research* 51, suppl. 7 (1992): 59–63.

72. Bjørn P. Kaltenborn, "Arctic-Alpine Environments and Tourism: Can Sustainability Be Planned? Lessons Learned on Svalbard," *Mountain Research and Development* 20, no. 1 (2000): 28–31.

73. R. L. Withington and L. W. Hall, "Snowmobile Accidents: A Review of Injuries Sustained in the Use of Snowmobiles in Northern New England during the 1968–1969 Season," *Journal of Trauma* 10, no. 9 (1970): 760–63.

74. Ibid.

75. M. Bauer and A. Hemborg, "Snowmobile Accidents in Northern Sweden," *Injury* 10, no. 3 (1979): 178–82.

76. A. Eriksson and U. Bjornstig, "Fatal Snowmobile Accidents in Northern Sweden," *Journal of Trauma* 22, no. 12 (1982): 977–82. Further study in Sweden revealed that 86% of those fatally injured in snowmobile accidents were driving under the influence of alcohol. See U. Bjornstig, "Accidents in the North: Some Aspects of Snowmobile Accidents and Moose-Car Collisions," *Arctic Medicine Research* 51, suppl. 7 (1992): 56–58.

77. M. Ostrom and A. Eriksson, "Snowmobile Fatalities: Aspects on Preventative Measures from a 25-Year Review," *Accident Analysis Prevention* 34, no. 4 (2002): 563–68.

78. G. S. Hortemo, G. Brattebo, and S. Hellesnes, "The Snowmobile—Only for Fun?" [in Norwegian], *Tidsskr Nor Laegeforen,* 110, no. 10 (1990): 1196–98. It is uncertain whether these accidents had something to do with the decrease in mean body temperature and in temperature at the extremities that accompanied snowmobile operation (which, researchers somehow discovered, was accompanied by slight increase in rectal temperature). See H. Virokannas and H. Anttonen, "Thermal Responses in the Body during Snowmobile Driving," *Arctic Medicine Research* 53, suppl. 3 (1994): 12–18.

79. C. R. Hamdy et al., "Snowmobile Injuries in Northern Newfoundland and Labrador: An 18-Year Review," *Journal of Trauma* 28, no.

8 (1988): 1232–37. Physicians examined 310 patients (237 of them males, 73% of them under thirty years of age, and 25.5% of them adolescents). Falls and collisions were the most common cause of injuries.

80. E. C. James et al., "Snowmobile Trauma: An Eleven-Year Experience," *American Surgeon* 57, no. 6 (1991): 349–53.

81. D. R. Fraley et al., "The Care and Cost of Snowmobile-Related Injuries," *Minnesota Medicine* 79, no. 12 (1996): 21–25.

82. Centers for Disease Control and Prevention, "Injuries and Deaths Associated with Use of Snowmobiles—Maine, 1991–1996," *Morbidity and Mortality Weekly Report* 46, no. 1 (1997): 1–4.

83. www.cdc.gov/mmwr/preview/mmwrhtml/00035502.htm.

84. Ibid. For the winter seasons of 1990–91 through 1995–96, a total of 1,355 snowmobile-related incidents were investigated. The incidents involved 1,828 snowmobiles and resulted in 903 injuries among 2,105 operators and passengers. A total of 1,565 (86%) of the snowmobile operators were male. Of the 1,740 operators for whom age was known, 1,076 (62%) were over twenty-five years old. Of the 1,817 operators and passengers for whom information was available, 1,312 (72%) were wearing helmets at the time of the incident.

85. Joseph Pierz, "Snowmobile Injuries in North America," *Clinical Orthopaedics and Related Research* 1, no. 409 (2003): 29–36.

86. Laura Aiken, "The Cold Truth about Snowmobiles," *Canadian Medical Association Journal* 168, no. 6 (2003): 753.

87. Maternal and Child Health Bureau of the Department of Health and Human Services, Rural Injury Prevention Resource Center, National Farm Medicine Center, "Snowmobiles and Youth Safety Packet: Injury Fact Sheet," March 1997, at www.marshmed.org/nfmc/projects/initiav/ProfEd/snowmobi.htm; American Academy of Pediatrics, "Snowmobile Statement," *Pediatrics* 82, no. 5 (1988): 798; Canadian Paediatric Society, "Two-, Three-, and Four-wheel Unlicensed Off-Road Vehicles," *Canadian Medical Association Journal* 136 (1987): 119–20.

88. Modified from the recommendations of the committee on injury and poison prevention of the American Academy of Pediatrics, "Snowmobiling Hazards," *Pediatrics* 106 (2000): 1142–44, at www.facs.org/trauma/snowmobile.html; "Snowmobiles and Youth Safety Packet: Injury Fact Sheet," at www.marshmed.org/nfmc/projects/initiav/ProfEd/snowmobi.htm; American Academy of Pediatrics, "Snowmobile Statement"; Canadian Paediatric Society, "Two-, Three-, and Four-wheel Unlicensed Off-Road Vehicles," 119–20; J. M. Decou et al., "Snowmobile Injuries and Fatalities in Children," *Journal of Pediatric Surgery* 38, no. 5 (2003): 784–87.

89. www.cpsc.gov/CPSCPUB/PREREL/prhtml81/81003.html.

90. www.snowmobilers.org/facts_soundmanagement.html (accessed July 14, 2005).

91. Ibid.

92. Ibid.

93. www.snowmobile.org/facts_sound.asp.

94. Reich, "Ski-Dogs, Pol-Cats, and the Mechanization of Winter," 499–502.

95. Bob Behme, "Those Liquid-Cooled Snowmobiles," *Mechanix Illustrated* 69 (December 1973): 30–31, 111.

96. Lodico, *Environmental Effects of Off-Road Vehicles,* 10–16.

97. Environmental Protection Agency, *Regulatory Announcement: Frequently Asked Questions from Snowmobile Owners,* EPA420-F-02-240 (Washington, DC: EPA, September 2002).

98. www.yamaha-motor.com. No longer on webpage.

99. www.offroad-training.org/. Honda has OHV (Off-Highway Vehicle) and Environmental Learning Centers building on existing Honda Rider Education Centers in Colton (CA), Troy (OH), Irving (TX), and Alpharetta (GA) that are intended to teach "proper operation of motorcycles and ATVs" and "overall responsible land use ethics."

100. *Polaris Post,* various, 1960s.

101. *Inside Tracks* 1, no. 5 (1993).

102. Samantha Booth (Miss Snowflake), email to author, July 20, 2005.

103. www.awsc.org/snowflake.htm.

104. Booth email, July 20, 2005.

105. Ibid.

106. www.snowmobilers.org/facts_statelaws.html.

107. Mark Okrant and Laurence Goss, "The Impact of Spending by Snowmobilers on New Hampshire's Economy during the 2002–03 Season," Plymouth State University, August 2003.

108. Office of Parks, Recreation, and Historical Preservation, State of New York, *2003 Snowmobile Survey* (Albany, 2003).

109. www.snowmobile.org/facts_econ.asp (accessed July 12, 2005).

110. Booth email, July 20, 2005.

111. www.snowmobile.org/pr_wildernesstruth1.asp.

112. Marshall, "Problem of Wilderness," 147.

113. For snow trail information, see www.snowmobile.org/facts_snowtrails.asp.

114. Sheila Gallagher, "Bluewater Sparks Snowmobile Debate," *Earth Island Journal,* fall 1999, p. 8.

115. James E. McCarthy, *Snowmobiles: Environmental Standards and Access to National Parks* (Washington, DC: Congressional Research Service, December 9, 2004).

116. Miguel Llanos at http://msnbc.com/news/836652.asp?ocv=CB10.

117. Michael Yochim, "Snowplanes, Snowcoaches, and Snowmobiles: The Decision to Allow Snowmobiles into Yellowstone National Park," *Annals of Wyoming* 70, no. 3 (1998): 7–13.

118. Ibid., 14.

119. Michael Yochim, "The Recent Winter Use History of Yellowstone National Park," *Annals of Wyoming* 73, no. 1 (2001): 34–36.

120. Ibid., 37–41.

121. Ibid., 37–41, and Gary Bishop, Jerome Morris, and Donald Stedman, "Snowmobile Contributions to Mobile Source Emissions in Yellowstone National Park," *Environmental Science and Technology* 35, no. 14 (2001): 2874–81. See also Rachel Petkewich, "Discord over Snowmobiles," *Environmental Science and Technology*, 35, no. 15 (2001): 318a–19a.

122. www.environmentaldefense.org/pressrelease.cfm?ContentID=2300 and www.google.com/u/envdef?q=patton+snow mobile&btnG=Search&hl=en&lr=&ie=ISO-8859-1.

123. Michael Janofsky, "U.S. Court Blocks Rules for Snowmobile Emission," *New York Times*, June 2, 2004, A15.

124. Letter from George B. Hartzog Jr. and others to Gale Norton, May 20, 2003.

125. *Congressional Record—House*, July 17, 2003, H7069, H7074, H7075.

126. Ibid.

127. Letter from Rush Holt, Christopher Shays et al. to Gale Norton, October 14, 2003. Fran Mainella, director of the National Park Service, answered by pointing out that the service had, in keeping with the National Environmental Protection Act, solicited more public comment, considered additional snowmobile technology, and taken into account information not available at the time of the original decision to phase out snowmobiles. The Park Service planned to reduce numbers of snowmobiles through limits, to require that snowmobiles use the best available technology to reduce emissions and noise, and to continue to monitor the situation. Letter from Fran Mainella to Rush Holt, November 12, 2003.

128. Thad Box, "Listening to the Land: Losses, Changes, and Teaching Moments," *Rangelands* 26, no. 6 (2004): 64–65.

129. Johnson interview, November 2, 2005.

130. Ibid.

131. Marco Musiani and Paul Paquet, "The Practices of Wolf Persecution, Protection, and Restoration in Canada and the United States," *BioScience* 54, no. 1 (2004): 50–60.

Chapter 3. Internal Combustion Adventure

1. http://4wheeldrive.about.com/gi/dynamic/offsite.htm?zi=1/XJ&sdn=4wheeldrive&zu=www.motorcycle-usa.com/Article_Page.aspx?ArticleID=174&Page=1.

2. www.atvconnection.com/atvconnection/Features/Editorial-Dummying-Down-The-Sport (accessed April 29, 2004).

3. www.geocities.com/missusamagicj/texas2004.html.

4. Izaak Walton League, *Caught in the Treads: Unethical Advertising in the ATV Industry* (Gaithersburg, MD: Izaak Walton League, 2002).

5. "Hill-and-Gully Riders," *Time* 93 (February 28, 1969): 90–91. Coot also built Snoopy, a mini–dune buggy capable of reaching 40 mph, powered by a Tecumseh 12-hp engine. The mini-buggy (at 7½ feet long and 4 feet wide) weighed 360 pounds and could carry 500 pounds. Snoopy was accompanied by the Searcher and the Red Baron dune buggies at the low and at the high end of the cost scale, respectively. Herbert Shuldiner, "'Snoopy' the Mini Dune Buggy," *Popular Science* 195 (December 1969): 50.

6. "Build MI's Fun Buggy!" *Mechanix Illustrated,* 69 (August 1973): 82–83.

7. Angus Laidlaw, "An ATV with Muscles," *Mechanix Illustrated* 65 (November 1969): 48–49.

8. "This Eagle Can Swim," *Mechanix Illustrated* 65 (December 1969): 63.

9. "Muckmobiles," *Hot Rod* 22 (May 1969): 120.

10. Bill Miller, "We Go ATV Camping," *Mechanix Illustrated* 69 (April 1973): 129–32.

11. Bill McAuliffe, *ATV Racing* (Mankato, MN: Capstone High/Low Books, 1999), 14–22.

12. www.atving.com/editor/feature/atvhistory/hondahistory.htm.

13. Ibid.

14. www.bombardier-atv.com/en-US/Index.htm, various models.

15. *Polaris Industries Inc. 2004 Annual Report,* 5–11.

16. www.atvsource.com/articles/press_releases/2005/050905_predator_500_troy_lee_designs_edition_wins_sport_quad_year.htm.

17. *Polaris Industries Inc. 2004 Annual Report,* 11. The relationship between military enterprise, innovation, and contracts and the civilian sector of the economy assumed a role in recreational vehicles too. In 1958 the U.S. Air Forces used the Polaris Sno-Trailer to get to within 400 miles of the North Pole.

18. On the relationship between control of nature and war, see Edmund Russell, *War and Nature* (Cambridge: Cambridge University Press, 2001).

19. www.comacclub.org/membership.asp.

20. www.bamaboggers.com (accessed December 8, 2005).

21. www.ozarktrailsatv.com/creed3.html.

22. http://4wheeldrive.about.com/gi/dynamic/offsite.htm?zi=1/XJ&sdn=4wheeldrive&zu=http%3A%2F%2Fwww.4wheelers4christ.org%2F4w4c%2FCalifornia.html.

23. www.state.me.us/doc/parks/programs/ATV/atv_club.html; ibid., atv_trailguide.html; ibid., atv_clubgrant.html; ibid., atv_plantrails.html.

24. "About NYSORVA," at www.nysorva.org.

25. Association for the Protection of the Adirondacks at www.protectadks.org/atv.html; www.dec.state.ny.us/website/environmentdec/2004b/atvenforce924.html.

26. www.dec.state.ny.us/website/environmentdec/2004b/atvenforce924.html.

27. Draft, ATV Policy for Public ATV Access to Recreational Programs, New York State Department of Environmental Conservation, March 9, 2005.

28. www.dec.state.ny.us/website/environmentdec/2005a/atvdraft031605.html.

29. Telephone interview with Peter Frank, July 26, 2005.

30. Ibid.

31. Ibid.

32. State of Maine, Department of Inland Fisheries and Wildlife, *All-Terrain Vehicles, 2001/02 Laws and Rules* (Augusta, 2001), 3.

33. Bureau of Parks and Recreation, Maine Department of Conservation, *A Report and Recommendations for the Management of All-Terrain Vehicles in Maine* (Augusta, February 1, 1986).

34. Herb Hartman, Department of Conservation, Department of Inland Fisheries and Wildlife, *Report on Maine's ATV Laws* (Augusta, January 31, 1989).

35. Roberta Scruggs, ATV Task Force coordinator, *ATV Solutions: Recommendations of Governor John Baldacci's ATV Task Force.*

36. www.state.me.us/ifw/rv/atvtaskforce/.

37. Peter Mosher, email to ATV Task Force, September 24, 2003, from files of Paul Jacques, deputy commissioner, Inland Fisheries and Wildlife Commission, State of Maine.

38. Lori Duniva, email to ATV task force, September 23, 2003, from files of Paul Jacques.

39. Colonel Timothy Peabody, chair, Law Enforcement Subcommittee of ATV Task Force, September 15, 2003, from files of Paul Jacques.

40. In Maine, no license is required to operate an ATV, but a person under age sixteen must complete a training program approved by

the department prior to operation. The new law prohibited a series of actions: operating an unregistered ATV; operating on a controlled-access highway; operating on a snowmobile trail financed in whole or in part by snowmobile funds; operating on a private road, public road, or sidewalk; failure to yield; reckless operation; DWI (a 0.8% blood alcohol level); operating to endanger; operating at faster than reasonable and prudent speeds; operating under eighteen years of age without headgear; operating on railroad tracks; driving too close to (within 200 feet of) a dwelling, hospital, nursing home, church, or convalescent home. ATVs must have a spark arrester and running lights and must be muffled so that they are no louder than 82 dB at 50 feet. They are prohibited in salt marshes, intertidal zones, marine sand beaches, sand dunes, cemeteries, alpine tundra, freshwater marshes, or bogs. See www.state.me.us/ifw/rv/atvlaws2003.htm (accessed June 4, 2003) and State of Maine, *All-Terrain Vehicles,* 6–11.

41. www.state.me.us/doc/parks/programs/ATV/atv_club.html; ibid., atv_trailguide.html; ibid., atv_clubgrant.html; ibid., atv_plantrails.html.

42. www.sec.state.vt.us/secdesk/opinions/2002/Oct02.html; www.vtfpr.org/recgrant/trgrant.cfm.

43. "Deaths among Drivers of Off-Road Vehicles after Collisions with Trail Gates—New Hampshire, 1997–2002," at www.cdc.gov.mmwr/preview/mmwrhtml/mm5203a2.htm (accessed June 11, 2003); www.nhparks.state.nh.us/trails/trails/trailspages/ATVpages/atvlaws.html; www.nhparks.state.nh.us/trails/trails/trailspages/GrantInAid.html, (both accessed June 4, 2003).

44. M. J. Wisdom et al., "Spatial Partitioning by Mule Deer and Elk in Relation to Traffic," in M. J. Wisdom, ed., *The Starkey Project: A Synthesis of Long-Term Studies of Elk and Mule Deer* (2005). Reprinted from the 2004 Transactions of the North American Wildlife and Natural Resources Conference (Lawrence, KS: Alliance Communications Group, 2004), 53–66.

45. www.fs.fed.us/eng/road_mgt/overview.shtml.

46. John Douglas Peine, "Land Management for Recreational Use of Off-Road Vehicles," Ph.D. diss., Department of Watershed Management, University of Arizona, Tucson (1972), 1, 3.

47. Kawasaki 2005 ATV brochures.

48. Robert Stebbins, "Off-Road Vehicles and the Fragile Desert," *American Biology Teacher,* May 1974, 294–304.

49. Geological Society of America, *Impacts and Management of Off-Road Vehicles: Report of the Committee on Environment and Public Policy* (Boulder, CO: GSA, May 1977), 1–5.

50. William J. Kockelman, "Introduction," in Robert H. Webb and

Howard G. Wilshire, *Environmental Effects of Off-Road Vehicles* (New York: Springer-Verlag, 1983), 1–5.

51. Ibid., 7.

52. Howard G. Wilshire, "The Impact of Vehicles on Desert Soil Stabilizers," in Webb and Wilshire, *Environmental Effects of Off-Road Vehicles,* 31–50; Earl Lathrop and Peter Rowlands, "Plant Ecology in Deserts: An Overview," in ibid., 113–38. On the slow recovery of soils and vegetation from human activity in ghost towns, see also Webb and Wilshire, "Recovery of Soils and Vegetation in a Mojave Desert Ghost Town, Nevada, USA," *Journal of Arid Environments* 3 (1980): 291–303.

53. Harold Dregne, "Soil and Soil Formation in Arid Regions," in Webb and Wilshire, *Environmental Effects of Off-Road Vehicles,* 15–26.

54. Geological Society of America, *Impacts and Management of Off-Road Vehicles,* 1–5.

55. Ibid., 6–7. See also Dale Gillette and John Adams, "Accelerated Wind Erosion and Prediction of Rates," in Webb and Wilshire, *Environmental Effects of Off-Road Vehicles,* 97, and Lathrop and Rowlands, "Plant Ecology in Deserts: An Overview," in ibid., 145.

56. Norma Jean Lodico, *Environmental Effects of Off-Road Vehicles: A Review of the Literature,* bibliography series no. 29 (U.S. Department of the Interior, Office of Library Services, September 1973), 10–16.

57. Bureau of Land Management, U.S. Department of the Interior, *Off-Road Vehicle Designations for Wilderness Study Areas in the Grass Creek Resource Area* (Worland, WY: BLM, 1990).

58. http://idaho.sierraclub.org/orv/#fact. One heavily used ORV site in the Los Padres National Forest has a documented soil loss exceeding 54,000 tons per square mile.

59. www.americanwilderness.org/reportcard/house_descriptions.htm.

60. www.leaveitwild.org/regions/utah/; www.americanwilderness.org/reportcard/house_descriptions.htm.

61. www.suwa.org/newsletters/1999/summer/scienc~1.htm.

62. T. M. Quigley and M. J. Wisdom, "The Starkey Project: Long-Term Research for Long-Term Management Solutions," in Wisdom, ed., *The Starkey Project,* 9–16.

63. B. K. John et al., "Elk and Mule Deer Responses to Variation in Hunting Pressure," in ibid., 127–38.

64. M. J. Wisdom et al., "Cattle and Elk Responses to Intensive Timber Harvest," 2005 in ibid., 197–216.

65. www.fs.fed.us/r6/wenatchee/recreate/recmain.html.

66. USDA Forest Service Environmental Statement, *Off Road-Vehicle Study,* USDA-FS-R6-DES-(Adm)-76-13 (Wenatchee National Forest: U.S. Forest Service, USDA, May 1976), 1–5.

67. Ibid., 27–32.

68. For archival materials pertaining to Udall's involvement in the commission, see http://dizzy.library.arizona.edu/branches/spc/udall/udallfindingaid/ufa/pllrc.htm.

69. On the Sagebrush Rebellion, see archival materials in the special collection at the University of Nevada, Reno, at www.library.unr.edu/specoll/mss/85-04.html.

70. Betsy A. Cody and Pamela Baldwin, "Bureau of Land Management Authorization," 95–429 ENR, March 27, 1995, at www.ncseonline.org/NLE/CRSreports/Natural/nrgen-7.cfm?&CFID=40312848&CFTOKEN=46709651.

71. Felicity Barringer, "In Utah, Trying to Undo a Federal Claim Bit by Bit," *New York Times,* November 24, 2005, A22.

72. General Accounting Office, Resources, Community, and Economic Development Division, *Federal Lands: Agencies Need to Assess the Impact of Personal Watercraft and Snowmobile Use,* GAO/RCED-00-243 (Washington, DC: U.S. GPO, 2000), 1–5.

73. Ibid.

74. Ibid.

75. Ibid.

76. Carol Greenberger, "The Controversy of ATVs on Public Lands," *Chattooga Quarterly,* at www.chattoogariver.org/Articles/2003Su/ATVs.htm.

77. www.friendsofthebitterroot.org/WSA_Watch.html.

78. "Off-Road Vehicles and Their Impact on Stream Environments: A Policy Statement from the Texas Chapter of the American Fisheries Society, January 2002," at www.sdafs.org/tcafs/content/orvpol.htm. For trade organization admonishments on ORV use, see www.suvone.com/suv415.htm, www.atvsafety.org/content/respectforoutdoors.html, and www.treadlightly.org.

79. www.atvconnection.com/atvconnection/Departments/Dunes/No-More-Junk-Science.cfm (accessed April 29, 2003).

80. Nancy Clemente, email to author, July 18, 2005.

81. "Program Update for Off-Road Motorcycles and ATVs," October 22, 1998, at www.arb.ca.gov/msprog/offroad/mcfactst.htm.

82. www.atvaonline.com/news/03/califnoise.asp. Wisconsin also accepted this standard according to Act 251 of 2003, which prohibited a person or manufacturer from selling, renting, or operating an ATV if the noise level exceeded 96 dB. In the ATV test, the ATV rests on a flat surface, usually gravel, in an area free of large reflective objects. It is straddled by a rider and placed in neutral with the brake held, and after a warm-up period a decibel meter is placed to the rear of the machine,

20 inches behind the exhaust at a 45-degree angle. The machine is revved to one-half the its maximum rpm. The snowmobile standard is 82 decibels, but the distance is 157 ½ inches from the sound source. http://66.102.7.104/search?q=cache:babkkHROOcMJ:www.dnr.state.wi.us/org/nrboard/minutes/M05/0505%2520minutes.pdf+decibels+atvs&hl=en.

83. Peter Demas and Thomas Braun, "Pediatric Facial Injuries Associated with All-Terrain Vehicles," *Journal of Oral Maxillofacial Surgery* 50 (1992): 1280–83; "CPSC Urges Caution for Three- and Four-Wheeled All-Terrain Vehicles," CPSC document 540 (Washington, DC: CPSC, 1987).

84. www.cdc.gov/mmwr/preview/mmwrhtml/00000524.htm.

85. J. Alexander Pyper and G. Brian Black, "Orthopaedic Injuries in Children Associated with the Use of Off-Road Vehicles," *Journal of Bone and Joint Surgery* 70-A, no. 2 (1988): 275–84.

86. www.cdc.gov/mmwr/preview/mmwrhtml/00056232.htm.

87. Demas and Braun, "Pediatric Facial Injuries."

88. Robin Ingle, *Annual Report of ATV Deaths and Injuries* (Washington, DC: CPSC, May 15, 2002).

89. Ronald Medford and Susan Ahmed, *Annual Report: All-Terrain Vehicle (ATV)–Related Deaths and Injuries,* CPSC memorandum, June 13, 2000.

90. Ibid.

91. Ingle, *Annual Report of ATV Deaths and Injuries.*

92. "Specialty Vehicle Institute of America Says ATV Safety Figures Not as Bad as Opponents Claim," March 8, 2004, at www.atvaonline.com/News/04/ATVfigures.asp.

93. Ibid.

94. National Safe Kids Campaign, *ATV Injury Fact Sheet* (Washington, DC: NSKC, 2004).

95. Honda and ATV Safety Institute, *Tips and Practice Guide for the ATV Rider* (2000), 11–23. All state ATV handbooks are based on this standard language used by all manufacturers.

96. www.atving.com/editor/feature/atvhistory/hondahistory.htm.

97. Committee on Government Operations, U.S. Congress, 100th Cong., 1st sess. (October 2, 1987), *Consumer Product Safety Commission's Response to Hazards of Three-Wheel All Terrain Vehicles (ATVs): A Follow Up Report* (Washington, DC: U.S. GPO, 1987), 1–5.

98. Ibid., 5–22.

99. Ibid., 23–25.

100. Committee on Governmental Affairs, U.S. Senate, *Regulation*

of All-Terrain Vehicles, 101st Cong., 2nd sess. (July 27, 1990), (Washington, DC: U.S. GPo, 1990), 1–3.

101. Ibid., 8.

102. Ibid., 11.

103. Ibid., 13–15.

104. Ibid., 50–52.

105. Ibid., 40–46.

106. www.cpsc.gov/CPSCPUB/PREREL/prhtml03/03136.html; memorandum, Hal Stratton, chairman, CPSC, to Patricia Semple, executive director, Consumer Federation of America, June 8, 2005; Rachel Weintraub, director, product safety, and assistant general counsel, CFA, email to author, August 22, 2005.

107. www.cpsc.gov/cpscpub/prerel/prhtml06/06004.html.

108. Greenberger, "Controversy of ATVs on Public Lands."

109. Dale Bosworth, "OHV Use: Rising to the Management Challenge," speech at ATV Expo industry breakfast, Louisville, KY, October 14, 2004.

110. Ibid.

111. "Over Three Hundred Concerned Citizens Organizations," letter to Mr. Dale Bosworth, chief, U.S. Forest Service, April 13, 2004.

112. See letter of Congressman James V. Hansen to George W. Bush and Richard Cheney, at www.npca.org/magazine/2001_issues/march_april/hansen.asp. Only three presidents since 1906 have failed to designate any such lands: Ronald Reagan, Richard Nixon (whose environmental record is still the envy of most presidents), and George Bush Sr. Bush Jr. recently designated the northwestern Hawaiian Islands a national monument to protect the waters and marine life around the 1,400-mile-long chain, but his actions regarding the national parks and other federal lands have drawn only criticism from environmental groups.

Chapter 4. The Lunacy of Personal Watercraft

1. David Evers, *Status Assessment and Conservation Plan for the Common Loon (Gavia immer) in North America* (Hadley, MA: U.S. Fish and Wildlife Service, 2004), 1, 4, 7–11, 24–25.

2. Catherine Ream, "Loon Productivity, Human Disturbance, and Pesticide Residues in Northern Minnesota," *Wilson Bulletin* 88, no. 3 (1976): 427–32. Study of New Hampshire lakes in the summers of 1976 and 1977 indicated that loons were leaving lakes probably because of loss of suitable nesting habitat, increased human disturbance, and increased depredation of nests. See S. A. Sutcliffe, "Changes in Status and Factors Affecting Common Loon Populations in New Hampshire," *Transactions*

of the Northeast Section, the Wildlife Society 35 (1978): 219–24. See also J. A. Munro, "Observations of the Loon in the Cariboo Parklands, British Columbia," *Audubon* 62 (1945): 38–49.

3. Evers, *Status Assessment and Conservation Plan for the Common Loon*, 1, 4, 7–11, 24–25.

4. Keith L. Bildstein et al., "Approaches to the Conservation of Coastal Wetlands in the Western Hemisphere," *Wilson Bulletin* 103, no. 2 (1991): 218–21.

5. www.dalesjetsports.com/watercraft/History_PWC.htm. See also http://web.mit.edu/invent/iow/watercraft.html.

6. With much less success than they calculated, engineers introduced fish for sporting and economic purposes. Because the dams slowed river flow, raised water temperature, and changed water chemistry, because decaying organic material in the new lakes did not behave as anticipated, and because newly introduced fish had no natural predators, the fisheries operated poorly.

7. Jane Pek, "Four If by Sea," *E Magazine*, September–October 2002, 24.

8. National Transportation Safety Board, *Personal Watercraft Safety* (Washington, DC: NTSB, 1998), 1–8.

9. As cited at www.kawasaki.ca/museum_pwc.html.

10. Mike Nixon, "The Personal Watercraft Phenomenon," at www.motorcycleproject.com/motorcycle/text/phenom3.html.

11. www.kawasaki.ca/museum_pwc.html.

12. http://web.mit.edu/invent/iow/watercraft.html.

13. www.kawasaki.ca/museum_pwc.html.

14. Frank Brown, "Power Play: Hatch a Wave and You're Sitting on Top of the World," *Crawdaddy*, September 1978, 14.

15. www.kawasaki.ca/museum_pwc.html.

16. Ibid.

17. Ibid.

18. Nixon, "Personal Watercraft Phenomenon"; www.seadoo.com/en-US/Watercrafts/About.Us/Sea-Doo.

19. www.seadoo.com/en-US/Watercrafts/About.Us/Sea-Doo/History/.

20. www.seadoo.com/en-US/Watercrafts/About.Us/Sea-Doo.

21. Personal Watercraft Industry Association, *The History, Evolution, and Profile of Personal Watercraft* (Washington, DC: PWIA, 2006), 2–3.

22. *Florida Fish and Wildlife News* 17, no. 3 (2003): 11. Thirty-seven percent of freshwater fish species are at risk of extinction, 35% of amphibians are rare or imperiled, and 67% of freshwater mussels are

rare or imperiled. Florida has fifty species of wildlife federally listed as endangered.

23. Telephone interview with Peggy Mathews, AWA, July 7, 2006.

24. James Rodgers Jr. and Stephen T. Schwiker, "Buffer-Zone Distances to Protect Foraging and Loafing Waterbirds from Disturbance by PWC and Outboard-Powered Boats," *Conservation Biology* 16, no. 1 (2002): 216–24. Rodgers's earlier work, intended to overcome such detriments to reproductive success as egg and nestling mortality, nest evacuation, reduced nestling body mass, premature fledging, and modified adult behavior, suggested set-back distances of about 100 m for wading birds and up to 180 m for mixed tern/skimmer colonies. See Rodgers and Henry T. Smith, "Set-Back Distance to Protect Nesting Bird Colonies from Human Disturbance in Florida," *Conservation Biology* 9, no. 1 (1995): 89–99, and Rodgers and Schwiker, "Buffer Zone Distances . . . Outboard-Powered Boats," 139–45. See also J. Burger and M. Gochfeld, "Human Distance and Birds: Tolerance and Response Distances of Resident and Migrant Species in India," *Environmental Conservation* 18 (1991): 158–65.

25. James Rodgers Jr. and Stephen Schwikert, "Buffer Zone Distances to Protect Foraging and Loafing Waterbirds from Disturbance by Airboats in Florida," *Waterbirds* 26, no. 4 (2003): 437–43.

26. Robin Jung, "Effects of Human Activities and Lake Characteristics on the Behavior and Breeding Success of Common Loons," *Passenger Pigeon* 53 (1991): 207–18.

27. Luanne Jaruzel, "Loon Watch," *Northern Michigan Journal,* 1998, at www.leelanau.com/njm/summer/loon.html.

28. Joe Brown, "Jet Skis Jolt Loons," *Audubon* 99 (September–October 1997): 14.

29. Jaruzel, "Loon Watch."

30. Raleigh Robertson and Nancy Flood, "Effects of Recreational Use of Shorelines on Breeding Bird Populations," *Canadian Field-Naturalist* 94 (1980): 131–38.

31. Marianne Heimberger, David Euler, and Jack Barr, "The Impact of Cottage Development on Common Loon Reproductive Success in Central Ontario," *Wilson Bulletin* 95, no. 3 (1983): 431–39.

32. *Impacts of Recreational Boating and PWC Use,* chap. 2, Commonwealth of Massachusetts PWC Management Guide, at http://64.233.187.104/search?q=cache:NrUhtzTAjgIJ:www.mass.gov/czm/pwcmgntguide2.pdf+impacts+of+recreational+boating+and+pwc+use&hl=en, 27–31.

33. www.pwia.org/issues/pwc_and_environment.html.

34. www.pwia.org/faqs/factsabout.html.

35. Mathews interview, July 7, 2006. See also Continental Shelf Associates, Inc., "Effects of Personal Watercraft Operation on Shallow Water Seagrass Communities in the Florida Keys," 1997.

36. www.conshelf.com/mission.html.

37. J. A. Rodgers Jr., *Buffer Zone Distances to Protect Foraging and Loafing Waterbirds from Disturbance by Personal Watercraft in Florida, Final Report* (Tallahassee: Florida Fish and Wildlife Conservation Commission, 2001).

38. Kevin Kenow, "A Voluntary Program to Curtail Boat Disturbance to Waterfowl during Migration," *Waterbirds* 26, no. 1 (2003): 77–87.

39. Ibid.

40. www.fws.gov/midwest/planning/uppermiss/#Next. According to the website, the Upper Mississippi Refuge includes "broad pools, islands, braided channels, extensive bottomland forest, floodplain marshes and occasional sand prairie. These habitats are critical to mammals, waterfowl, songbirds and raptors, amphibians and reptiles. More than 130 Bald Eagle nests and a yearly average of 14 active Heron colonies with a total of 5,000 nests exist on the Refuge. The Refuge and the River support 119 fish species that support a strong commercial and recreational fishery."

41. Kenow, "Voluntary Program to Curtail Boat Disturbance."

42. Ibid.

43. www.fws.gov/midwest/planning/uppermiss/DraftEIS.html.

44. www.awahq.org/p/newsDetail.php?entryID=101&flagger=1.

45. Mathews interview, July 7, 2006.

46. www.pwia.org/issues/pwc_and_environment.html.

47. www.pwia.org/faqs/factsabout.html; *Impacts of Recreational Boating and PWC Use,* chap. 2, 23; www.rbbi.com/company/omc/polaris.htm.

48. U.S. Coast Guard Media Advisory, April 2, 2001, at www.uscg.mil/news/Headquarters/Evinruderecall.htm. See also www.house.gov/transportation/cgmt/05-15-01/05-15-01memo.html#BACKGROUND.

49. www.findarticles.com/p/articles/mi_moBQK/is_3_6/ai_74699547.

50. www.sportfishingmag.com/article.jsp?ID=3306. In February 1998 the Outboard Marine Corporation "signed an agreement to license its proprietary low-emission FICHT fuel injection technology to Polaris Industries" for use on Polaris PWCs, snowmobiles, and ATVs. See www.rbbi.com/company/omc/polaris.htm.

51. www.epa.gov/otaq/regs/nonroad/marine/420f04031.pdf; www.epa.gov/otaq/cleaner-nonroad/f03011.pdf. See also EPA, *Overview of EPA's Emission Standards for Marine Engines,* EPA420-F-04-031, August

2004. Noxious engine emissions were not the only problem with marine engines. Fuel tank and other evaporative emissions fouled the air. In the absence of low-emission fuel tanks and evaporative controls, the EPA determined that over 100,000 tons of gasoline vapors were released across the United States annually. The emissions occurred either from permeation of the fuel through plastic fuel tanks and rubber hoses or as a result of heating of fuel from normal daily temperature changes. The vapors contributed to the formation of smog and contained such toxic, and in some cases carcinogenic, compounds as benzene. The EPA proposed an 80% reduction on current marine vessels.

52. Laurie Martin, *Caught in the Wake* (Gaithersburg, MD: Izaak Walton League of America, 1999), 6–12.

53. Stephanie Nowacek, Randall Wells, and Andrew Solow, "Short-Term Effects of Boat Traffic on Bottlenose Dolphins . . . in Sarasota Bay, Florida," *Marine Mammal Science* 17, no. 4 (2001): 673–88.

54. "Air Board Acts to Reduce Marine Engine Pollution," release 98–75, December 10, 1998, at www.arb.ca.gov/newsrel/nr121098.htm.

55. "First Personal Watercraft Certified to ARB 2001 Standards," release 99–12, April 13, 1999, at www.arb.ca.gov/newsrel/nr041399.htm; EPA, *Effect of Proposed Evaporative Emission Standards for Marine Manufacturers,* EPA420-F-02-008, July 2002. Polaris's two-stroke 122-hp 1.2-liter fuel-injected engine was three to four times cleaner than a conventional two-stroke PWC engine. EPA tests on the Genesis certified it at 44.28 grams per kilowatt hour of the smog-forming emissions hydrocarbons (HC) and oxides of nitrogen (NOx), while California's approaching 2001 emission standard for a watercraft engine of this horsepower was 46.15 HC plus NOx.

56. Shirley Land, "Boat Dealers Decry Plan on Emissions," *Wall Street Journal,* December 2, 1998, CA1. Automobile manufacturers too had predicted economic disaster when California set emissions standards higher than the federal ones. They too were proven wrong. California automobile emissions laws that were stricter than the national requirements enabled the manufacturers to produce better, cleaner engines.

57. Charles Komanoff and Howard Shaw, *Drowning in Noise: Noise Costs of Jet Skis in America,* report to Noise Pollution Clearinghouse, Vermont, April 2000. Komanoff and Shaw established the noise costs of jet ski operation using well-established and accepted techniques as a foundation for their analysis. When they plugged in empirical values (the jet ski is so many feet offshore, it is so many decibels loud at the source, the beach is so many feet wide and deep, the beach is "popular" or "secluded" with a corresponding background noise level, the average beachgoer spends so many dollars to be at the beach), the model yielded

estimates of the additional decibels of jet ski noise, the value of people's time at the beach, and the dollar value (the cost to them) of their reduced pleasure. Translating noise into dollars' worth of "disamenity" is nothing new.

58. At www.nonoise.org/library/drowning/drowning.htm.

59. www.pwia.org/faqs/factsabout.html.

60. Personal Watercraft Industry Association, *History, Evolution, and Profile of Personal Watercraft,* 10.

61. The PWIA cites a 1999 noise study conducted by the Finnish Ministry of the Environment that tested the noise levels of old and new PWCs and compared them with the noise levels of boats with two-stroke and four-stroke outboard engines and of a cabin cruiser with a stern-drive engine. The noise of the various boats was measured at different speeds, at different distances, and by irritation to observers, a subjective measurement. The Finnish researchers found that the new PWC was the quietest and least disturbing boat at speeds up to 40 kph in comparison with other types of boat tested. The new PWC was much quieter than the old PWC in every test, by an average 3.6 dB (comparing all speeds and distances). Considering that the decibel range is exponential, this is a significant reduction in sound output. See www.pwia.org/faqs/factsabout.html.

62. Sean Smith and Katy Rexford, *Personal Watercraft Production/Design Problems: High Potential for Fires and Explosions* (San Francisco: Bluewater Network, October 15, 2001).

63. Sean Smith and Carl Schneebeck, *Personal Watercraft Production/Design Problems: High Potential for Fires and Explosions* (San Francisco: Bluewater Network, March 2004). See, for example, Kawasaki, *JS-99-05,* December 17, 1999, and *JS-00-02,* October 26, 2000.

64. Amy Fox and Ellinore Boeke, "The Facts about Personal Watercraft Recalls," October 16, 2001, at www.pwia.org/news/news10.html.

65. National Transportation Safety Board, *Personal Watercraft Safety,* 9, 27.

66. Ibid., 10.

67. Ibid., 1, 8, 17. In one case a twenty-five-year-old male lost control of a rented PWC at Four Bear Water Park near Shelby Township, Michigan. He entered a lifeguard-supervised swimming area and struck six children between the ages of five and twelve, all of whom were injured, though none fatally.

68. Ibid., 17–22, 33, 36.

69. *Impacts of Recreational Boating and PWC Use,* chap. 2, 7–8.

70. National Transportation Safety Board, *Personal Watercraft Safety,* 17–22, 33, 36.

71. U.S. Coast Guard, U.S. Department of Transportation, *Boating Statistics—2000*, Comdtpub. P16754.14 (Washington, DC: USCG, 2001).

72. *Impacts of Recreational Boating and PWC Use,* chap. 2, 8–10.

73. "Drowning in Noise," at www.nonoise.org/library/drowning/drowning.htm, 56 (accessed July 27, 2005). Since Coast Guard data record only the craft used by the deceased, the PWC fatality figure omits cases in which a PWC caused the death of someone other than the user.

74. The following information about industry response to criticism of PWC safety is drawn from www.pwia.org/faqs/factsabout.html.

75. U.S. Coast Guard, U.S. Department of Transportation, *Boating Statistics—2000.*

76. *Florida Fish and Wildlife News* 17, no. 3 (2003): 11; www.floridaconservation.org/law/boating/2003stats/summary.htm.

77. B. L. Hamman, F. B. Miller, M. E. Fallat, and J. D. Richardson, "Injuries Resulting from Personal Watercraft," *Journal of Pediatric Surgery* 28, no. 7 (1993): 920–22. Children were often involved in accidents that required surgical intervention and that physicians believed were more serious than those involving other watercraft. See Elizabeth Beierle et al., "Small Watercraft Injuries in Children," *American Surgeon* 68 (June 2002): 535–38.

78. R. A. Francis and R. Vize, "Personal Watercraft Injuries: Experience at a Community Hospital," *Journal of Missouri Medicine* 91, no. 5 (1994): 241–43.

79. C. M. Branche, J. M. Conn, and J. L. Annest, "Personal Watercraft–Related Injuries: A Growing Public Health Concern," *Journal of the American Medical Association* 278, no. 8 (1997): 663–65.

80. D. V. Shatz et al., "Personal Watercraft Injuries: An Emerging Problem," *Journal of Trauma* 44, no. 1 (1998): 198–201. See also R. Latch and D. H. Fiser, "The Increasing Threat of PWC Injuries," *Clinical Pediatrics* 43, no. 4 (2004): 309–11, which expressed concern about both the increasing number of injuries and their increasing severity.

81. C. S. Jones, "Drowning among Personal Watercraft Passengers: The Ability of PFDs to Preserve Life on Arkansas Waterways, 1994–1997," *Journal of the Arkansas Medical Society* 96, no. 3 (1999): 7–98; C. S. Jones, "Epidemiology of PWC-Related Injury on Arkansas Waterways, 1994–1997," *Accident Analysis Prevention* 32, no. 3 (2000): 373–76.

82. J. I. Garri et al., "Patterns of Maxillofacial Injuries in Powered Watercraft Collisions," *Plastic Reconstructive Surgery* 104, no. 4 (1999): 922–27.

83. J. M. Philpott et al., "Rectal Blowout by Personal Watercraft Water Jet: Case Report and Review of Literature," *Journal of Trauma* 47, no. 2 (1999): 385–88; B. Descottes et al., "Rectal Injury Caused by Personal Watercraft Accident: Report of a Case," *Diseases of the Colon and Rectum* 46, no. 7 (2003): 971–73.

84. A. Tsai, J. T. Rhea, and R. A. Novelline, "The Jet Ski Open-Book Pelvic Fracture," *Emergency Radiology* 10, no. 2 (2003): 96–98.

85. J. M. Haan, M. E. Kramer, and T. M. Scalea, "Pattern of Injury from Personal Watercraft," *American Surgeon* 68, no. 7 (2002): 624–27. See also A. Carmel et al., "Thoracolumbar Fractures Associated with the Use of PWC," *Journal of Trauma* 57, no. (2004): 1308–10.

86. Mathews interview, July 7, 2006.

87. National Transportation Safety Board, *Personal Watercraft Safety*, 8.

88. http://archives.charleston.net/org/ccjsc/news.htm.

89. www.spacecoastjetriders.com/about.htm.

90. www.pwia.org/issues/watercraftcodeofethics.html.

91. *Impacts of Recreational Boating and PWC Use*, chap. 2, 20–22.

92. www.mecreative.com/pwc/friends.htm.

93. Jetty Jumpers of Coney Island, Brooklyn, NY, at www.jetty jumpers.com.

94. *Citizens for Florida's Waterways*, newsletter 10, no. 2 (2004): 9.

95. Josh Clemens, research counsel, Mississippi-Alabama Sea Grant Legal Program, letter to Dave Burrage, extension professor of marine resources, MSU Coastal Research and Extension Center, May 25, 2004.

96. "Stop Spread of Jet Skis to Niobrara National Recreational River," *Bluewater Network*, June 22, 2000, www.earthisland.org/ takeaction/actionalert_bw27.html.

97. Ibid.

98. Clemens to Burrage, May 25, 2004. The Administrative Procedure Act provided a legal basis for Bluewater to sue for administrative agency action and to have the prior action be set aside if it was found to be "arbitrary, capricious, an abuse of discretion or otherwise not in accordance with law."

99. Ibid.

100. Kristin Young, "Attack on the Water: The Facts about the Personal Watercraft Ban in Our National Parks," March 2001, at www .pwia.org/news2001/news40.html.

101. PWIA, "Lawsuit Seeks Delay of Personal Watercraft Ban to Allow National Parks to Complete Environmental Studies," March 28, 2002, at www.pwia.org/news2002/032802.html.

102. Young, "Attack on the Water."

103. PWIA, "Lawsuit Seeks Delay."

104. Ibid.

105. Clemens to Burrage, May 25, 2004.

106. "PWCs Banned from Most of Park System," *NPCA News and Notes,* July–August 2002, 14.

107. Clemens to Burrage, May 25, 2004.

108. Josh Clemens, email to author, October 11, 2005.

109. www.southernshores.org/Download%20Docs/PWC%20flyer.pdf.

110. Ibid.; www.nonoise.org/resource/jetskis/jsmemo.htm.

111. Martin, *Caught in the Wake,* 4.

112. www.npca.org/magazine/2002/april_may/news5.asp.

113. www.ncsl.org/programs/transportation/99boatlg.htm.

114. www.state.ny.us/governor/press/05/july28_05.htm.

115. www.pwia.org/news2005/050405.html.

116. Ibid.

117. "Personal Watercraft Industry Challenges Keys PWC Ban: Florida DEP to Review a 10-Year Jet Ski Ban in Lower Keys Wildlife Refuges," at www.floridasportsman.com/casts/050204e/.

118. Personal Watercraft Industry Association, *History, Evolution, and Profile of Personal Watercraft,* 1–9.

119. www.michiganloons.org/watercraft.htm.

Chapter 5. Small-Bore Engines around the Home and Garden

1. F. Herbert Bormann, *Redesigning the American Lawn: A Search for Environmental Harmony* (New Haven: Yale University Press, 2001), 46–65. For historical background, see Theodore Steinberg, *American Green: The Obsessive Quest for the Perfect Lawn* (New York: W. W. Norton, 2006).

2. "Hundred Years from the Scythe," *Garden Magazine* 37 (June 1923): 268–69.

3. Fred Haxton, "Sheep as Lawn Mowers," *Country Life in America* 11 (April 1907): 681.

4. "Gasoline-Operated Railway Weed Mower," *Scientific American* 113 (October 23, 1915): 384.

5. Leonard Barron, "Motorizing the Lawn Mower," *Garden Magazine and Home Builder* 41 (March 1925): 56–57. Some of the manufacturers in the mid-1920s were Acme Cultivator, Federal Foundry Supply, Philadelphia Lawn Mower, Jacobsen, Gilson, American Farm Machine, Caldwell Lawn Mower, Moto-Mower, and S. P. Townsend.

6. www.anla.org/applications/PressReleases/releases/0122.htm (accessed January 1, 2004).

7. Stihl claims to be "the leader" in environmental protection, a

claim that is "not merely a matter of theory." Work stations are designed to protect employees from noise and air pollution. Liquids, particles, and other by-products of manufacturing are constantly filtered out of water and air. Stihl recycles, makes minimal use of packing and shipping materials, and uses cadmium-free plastics and lead-free materials in the manufacturing process. See Stihl, "Welcome to Our Neck of the Woods," corporate brochure, Virginia Beach, VA, n.d. [2005?].

8. www.opei.org/about/history.asp.

9. Bormann, *Redesigning the American Lawn,* 66–80.

10. www.goatsrus.com.

11. Jim Killeen, "Nylon-Line Trimmer Industry Booming as New Products Enter Market Place," *Flower and Garden* 23 (May 1979): 44–45, 62; "Guide to Power Trimmers and Edgers," *Mechanix Illustrated* 64 (July 1968): 88–89; Gless Hensley, "New Ways with Nylon-Cord Lawn Trimmers," *Popular Mechanics* 152 (August 1979): 162–65.

12. Quoted in Hensley, "New Ways with Nylon-Cord Lawn Trimmers."

13. Killeen, "Nylon-Line Trimmer Industry Booming."

14. www.nonoise.org/quietnet/cqs/leafblow.htm.

15. Ibid.

16. Ibid.

17. Ibid.

18. Richard Estrada, "Political and Social Issues Accompany Leaf Blower Controversies in U.S.," *Dallas Morning News,* September 14, 1998, B6, as cited at www.nonoise.org/news/1998/sep13.htm.

19. "Policy Report: Environment," June 15, 2001, at www.city.vancouver.bc.ca/ctyclerk/cclerk/010712/csb5.htm.

20. Estrada, "Political and Social Issues."

21. www.clca.org/gov_blowers.html.

22. Said one environmental lawyer, "We Americans love our lawns and treasure our power tools. Our lawn mowers, trimmers, hedge clippers, snow blowers. The king of our power equipment arsenal is undoubtedly our leaf blowers. They are large, gas driven power machines. Size counts, and the bigger the leaf blower, the more leaves that can be forced onto the curb in the least amount of time. And Americans do not have that much time to devote to their leaves, so leaf blowers are vital to our existence. Most Americans treasure their inalienable right to bear leaf blowers. And it is this very right which is now under a nationwide attack. You better watch out: if you are not careful you may have to revert to raking. Perish that thought. Case in point: Los Angeles, where last year that City enacted an ordinance that banned leaf blowers all together." Stuart Lieberman, "On The Right to

Bear Leaf Blowers," at http://realtytimes.com/rtnews/rtcpages/19981124_leafblower.htm.

23. EPA, *Small Engine Emissions Standards,* EPA420-F-98-025 (Ann Arbor, MI: EPA National Vehicle and Fuel Emissions Laboratory, August 1998); EPA, *Your Yard and Clean Air,* fact sheet OMS-19 (Ann Arbor, MI: EPA National Vehicle and Fuel Emissions Laboratory, 1996).

24. www.cdc.gov/niosh/93-119.html.

25. Consumer Product Safety Commission, "Power Lawnmower Injuries Decline," release 90–131 (Washington, DC: CPSC, July 1990).

26. Consumer Product Safety Commission, "Vote for Safety Standard for Walk-Behind Power Lawn Mowers," release 79-006 (Washington, DC: CPSC, 1979).

27. www.washington.edu/newsroom/news/2002archive/06-02archive/k062702a.html. Each year in the United States approximately ninety-four hundred children under the age of eighteen receive emergency care for lawn mower–related injuries. Although most of these injuries occur to older children and adolescents, about one-fourth are to children under the age of five. Males account for approximately three-fourths of these injuries. Ride-on mowers and other power mowers account for 21% and 23% of pediatric mower-related injuries, respectively. More than 7% of pediatric mower-related injuries require hospitalization, which is approximately twice the hospitalization rate for consumer product–related injuries overall. Amputations and avulsions account for 7% of pediatric mower-related injuries. Power lawn mowers caused 22% of the amputation injuries among children admitted to one regional level 1 trauma center. See Committee on Injury and Poison Prevention, "Lawn Mower–Related Injuries to Children," *Pediatrics* 107, no. 6 (2001): 1480–81, and R. T. Loder et al., "Extremity Lawn Mower Injuries in Children: Report by the Research Committee of the Pediatric Orthopaedic Society of North America," *Journal of Pediatric Orthopaedics* 17, no. 3 (1997): 360–69. These horrible trends continue. See D. Vollman et al., "Lawn Mower–Related Injuries to Children," *Journal of Trauma* 59, no. 3 (2005): 724–28.

28. Consumer Product Safety Commission, "CPSC-Industry to Develop Safety Rules for Chain Saws," release 78-019 (Washington, DC: CPSC, March 1978).

29. Consumer Product Safety Commission, "Commission Approves Mandatory Approach toward Reducing Chain Saw Injuries from 'Kickback,'" release 80-023 (Washington, DC: CPSC, June 1980); Consumer Product Safety Commission, "Chain Saw Kickback Significantly Reduced," release 85-038 (Washington, DC: CPSC, August 1985).

30. Consumer Product Safety Commission, "CPSC and Husqvarna Announce the Recall of Chain Saws," release 97-098 (Washington, DC: CPSC, April 8, 1997).

31. Rachel Carson, *Silent Spring* (Boston: Houghton Mifflin, 1962), 5–6.

32. Marc J. Mazerolle, "Amphibian Road Mortality in Response to Nightly Variations in Traffic Intensity," *Herpetologica* 60, no. 1 (2004): 45–53. A subsequent study confirmed the danger of motorized vehicles to frogs, toads, and salamanders. See Mazerolle et al., "Behavior of Amphibians on the Road in Response to Car Traffic," *Herpetologica* 61, no. 4 (2005): 380–88.

INDEX